THE PROBLEM
OF
SCIENTIFIC REALISM

CONTEMPORARY PROBLEMS IN PHILOSOPHY
George F. McLean, O.M.I., *Editor*

Situationism and the New Morality, *Robert L. Cunningham*
Human and Artificial Intelligence, *Frederick J. Crosson*
The Problem of Scientific Realism, *Edward A. MacKinnon*

THE PROBLEM
OF
SCIENTIFIC REALISM

EDWARD A. MAC KINNON
California State University

New York
APPLETON-CENTURY-CROFTS
Educational Division
MEREDITH CORPORATION

CONTENTS

PREFACE

This book is intended for students of philosophy of science and for all who are interested in the historical or philosophical aspects of the problem of scientific realism. While the book presupposes some background in philosophy and some familiarity with the basic problems and schools of thought in the philosophy of science, it does not presuppose any particular philosophical orientation.

The long Introduction has two aims. First, it presents a historical survey focused on the theories that philosophers and reflective scientists have developed on the nature of scientific explanation. This survey is selective rather than complete. I was more concerned with using historical positions to bring out the complexity of the problem than with presenting history for its own sake. In particular I hoped to bring out the intimate relation between theories of knowledge and philosophical positions defending or attacking a doctrine of scientific realism.

The second half of the Introduction presents my evaluation of the problem and of the various theories that have been considered. As the historical survey indicates, an epistemological orientation is often the controlling element in any philosopher's appraisal of the problem of scientific realism. While I attempt to preserve a certain degree of philosophical objectivity, I cannot pretend that the appraisal offered here is not conditioned by my own epistemological position. Although I was originally trained as a Thomist, my present approach to philosophy is more problem-centered than doctrine-centered. The methodology followed draws heavily on the analytic and pragmatic traditions, while the evaluation of scientific realism is closer to Kantianism than to any other philosophical tradition. This was not a position adopted *a priori*, but one that resulted from an analysis of the arguments presently available to support metaphysical systems.

The selections included were chosen to bring out different schools of thought and different aspects of the problem. Except for the excerpts from Aristotle and Newton, which, I believe, are indispensable in understanding the historical dimensions of the problem, the authors selected are all contemporary writers influential in the

English-speaking world. The selections were chosen as units that
are intelligible in themselves. Some of them are rather difficult, but
none of them require a technical background in logic or science.
The supplementary bibliographical essay is intended as a guide for
further reading.

I wish to thank the Reverend George F. McLean, O.M.I., for the
invitation to contribute to this series and for his editorial assistance
in preparing the manuscript. I am grateful to Mrs. Louise Dieten-
hofer and Mrs. Marie Allen for typing the final draft of the Introduction.

E. MacK.

INTRODUCTION

Historical Development of Scientific Realism

Abiding Influence of Athens

Plato and Aristotle witnessed and participated in the burgeoning of science as a distinctive form of knowledge. Mathematics and astronomy had been developed in Egypt and Mesopotamia as practical rather than explanatory disciplines. There, explanation had been relegated to the province of mythology and its priestly interpreters, rather than to the clerks and computers who made calculations. In Greece, on the contrary, Thales, Pythagoras, and their successors believed that nature must be intelligible in its own terms, rather than in terms of anthropomorphic projections of human traits into the forces of nature. From their speculations emerged both science and philosophy as we now know them. Plato and Aristotle interrelated these disciplines through theoretical frameworks that dominated two thousand years of development and interpretation and still exert a strong influence. Both were realists, but interpreted realism quite differently. We wish here to clarify the significance each attached to scientific realism and to outline some aspects of their thought pertaining to the contemporary problem of scientific realism.

Plato's theory of knowledge and the closely related problem of the ontological status accorded platonic forms has supplied material for an unending debate among interpreters. Yet some basic aspects of his theory are beyond dispute. True knowledge must possess two qualities: it must be certain; and it must be knowledge of what is. Sensation fulfills neither criterion. One desiring candidates for the

3

lofty status of true knowledge must look to universals and definitions. Universals, i.e., terms like 'man' that do not denote a particular subject the way a proper name does, can fulfill these criteria
of being certain and being knowledge of what is only if they have
objective reference. For Plato these referents were the forms,
although even he was not altogether consistent about their precise
status. By telescoping Plato and later philosophers influenced by
him, we may list four possible, though not mutually consistent,
interpretations of platonic forms.

First, the forms could be thought of as real beings existing in
such a way that other existents may participate in these forms. In
this interpretation, influential in later neoplatonism and Augustinian-
ism, there is no sharp exclusion of non-members, but there is an
ordering according to degrees of participation. Secondly, forms
could be considered statically present in individuals, for example,
a form as a *class* of all members having a given property. In this
case there is no ordering, but there is a sharp division between
members and non-members, such as that between odd and even
numbers. Thirdly, according to an interpretation adumbrated in
the *Timaeus* and developed by Whitehead, a form could be considered as a process gradually evolving toward the perfection signified by its definition. Finally, a form could be thought of as a
criterion for class membership. In this case, unlike the second, the
form of a class is not a member of the class. The second and fourth
interpretations have had a checkered history in contemporary logic.
Regardless of the particular interpretation adopted, Plato's basic
emphasis remains a challenge. Unless universals have objective reference to what is, no true knowledge is possible.

Plato's interpretation of scientific knowledge was complicated by
the developments and controversies in which he participated. The
School of Megara, under Zeno, stressed logic and relied on paradox
and *reductio ad absurdum* arguments to deflate metaphysical pre-
tensions. Something of the same stress on formal methods had
entered Plato's Academy with the admission of Eudoxos, a great
mathematician and founder of the axiomatic method. To be adequate to these developments, Plato's theory of knowledge had to
include an account of science considered as a formal system, rather
than as a mere collection of statements.

In the image of the divided line,[1] Plato schematized a theory of knowledge that came to grips with these problems. The lower section of the line represents the world of appearances. Here one has, not real knowledge, but conjecture and opinion. The upper portion of the line, representing the intelligible order, is divided into two unequal sections. The lower and shorter section represents deductive systems of knowledge, knowledge that cannot transcend the assumptions from which its conclusions are derived. The objects appropriate to this form of knowledge are mathematical objects and functional relations. The highest form of knowledge, represented by the topmost part of the line, advances from assumptions to principles that transcend these assumptions or any dependence on images. Ideally it terminates in intuitive knowledge of pure forms.

Here, in germinal form, is a hierarchical ordering of types of knowledge and their correlative objects accepted and elaborated by later thinkers. Conjecture is subordinate to belief, belief to scientific knowledge, and science to metaphysics or to a non-deductive knowledge of first principles. Then, as later, many refused to accept the idea that science can be rightfully and realistically interpreted only by subordinating it to metaphysics. One may argue that the purely formal method of logical deduction from assumptions represents the highest form of knowledge. Plato considered such objections in his later dialogues. Thus the Parmenides, according to Brumbaugh's interpretation,[2] demonstrated by a *reductio ad absurdum* argument that formal systems, or the knowledge proper to the third level of the divided line, might be self-consistent yet mutually contradictory. The implication was that formal systems were valid vehicles of knowledge only if they were anchored in, and interpreted through, a higher form of knowledge, later called metaphysics.

Aristotle founded, in Whitehead's terms, the first physical synthesis, a synthesis that dominated interpretations of the scientific enterprise for two thousand years. Yet, as Jaeger has stressed, any attempt to extract a coherent synthesis from Aristotle's work to some degree misinterprets him. His thought changed and developed so that his later practice, which was highly empirical, did not always

1. Plato, *Republic,* 6, 509d–511e.
2. Robert Brumbaugh, *Plato on the One: The Hypothesis in the Parmenides* (New Haven: Yale Univ. Press, 1961).

accord with his earlier theories of scientific practice. The schematic outline that follows, though based on Aristotle's writings, could more appropriately be considered as a presentation of Aristotelianism than of Aristotle himself. We shall first examine Aristotle's theory of science, particularly his interpretation of scientific realism, and then his scientific practice.

In his *Posterior Analytics,* part of which is included as the first selection of the readings, Aristotle presents his interpretation of the structure and foundations of science. For the scientist as well as the student, the starting point is pre-existent knowledge, the ordinary knowledge one has of things and of the meanings of words. The scientist or natural philosopher differentiates one part of this knowledge, analyzes, orders, and develops it. There are four general requirements in the constitution of any science. The first is a genus, or general class, proper to the subject under consideration. The second is attributes that may be predicated of these subjects. The goal of scientific explanation is to determine which attributes are predicated of a subject essentially.

The third requirement, one that must be treated in more detail, is axioms or principles from which conclusions can be deduced. These axioms should fulfill strict specifications. They must be true, primary in the sense of being basic truths relative to the subject matter, certain, and of the same nature as the conclusion. In this final specification Aristotle is at variance both with his predecessors, the atomists, and with his eventual successors, the Newtonians. The former tried to explain perceptible bodies in terms of imperceptible atoms; the latter spoke of deriving physical conclusions from mathematical principles. Neither accords with Aristotle's idea that the axioms and conclusions must be of the same nature.

The final requirement is a deductive process leading from principles to conclusions. Here Aristotle relied, at least to some degree, on the syllogistic logic he had invented and systematized. If all *A* is *B* and all *C* is *A*, then all *C* is *B*. The middle term, *A*, not only links the subject and predicate, but also ideally gives the reason why the attribute is predicated of the subject. Although Aristotle's logic was a purely formal tool, the ontic commitment involved in its employment was a world of objects that could be classified in species, or types, and genera, or higher species, and of which attributes could be predicated. In "man is a rational animal," 'man' names a species, 'animal' a genus, and 'rational' an attribute specific to man.

As a convenient way of schematizing the different views to be considered, one could think of a scientific theory as a deductively ordered set of statements running from axioms at the top of the page, through the theorems, or general conclusions, in the middle of the page, to the particular conclusions at the bottom of the page. Aristotle's is strictly a top-of-the-page interpretation of scientific explanation. The conclusions are true and certain *because* the axioms are, and because the deductive process leading from axioms to conclusions is truth-preserving. The crucial question, accordingly, is how does one attain such axioms?

Since an infinite regress is inadmissible, explanation must begin with axioms whose truth is known in a non-deductive way. These axioms are of two sorts: general axioms, such as "equals subtracted from equals, yield equals," which may be used in any science, and proper axioms, which state primary truths about the subject matter in question. The latter, as Aristotle saw it, are attained by analysis, where 'analysis' has a broader and vaguer meaning than in current philosophy. The goal Aristotle aspired to was an analysis of beings into their causes. His example illustrates his classification of the four causes. The *material* cause of a statue is the marble, while the particular shape, e.g., Zeus, is its *formal* cause. The work of the sculptor imposing form on matter is the *efficient* cause, while the purpose for which it was carved is the *final* cause. Aristotle tried to extend this categorization from artificial to natural beings chiefly through his doctrine of substantial form. A thing is a particular type of being by virtue of a substantial form, which is the same for any member of a species. This identity of form within a species is the basis for the universality proper to scientific statements.

This search for causes and interpretation through causes formed the core of Aristotle's doctrine of scientific realism. If the expression of a cause serves as a middle term justifying the predication of an attribute to a subject, then one has both a certainty that the conclusion is true and an understanding of why it is true. The axioms, in turn, are true, primary, and certain because they express a causal knowledge.

In practice Aristotle's search for principles proper to the different sciences involved dialectical, linguistic, and phenomenological anal-

ysis, though Aristotle neither used these modern terms nor dis-
entangled the processes. The excerpts reproduced here indicate
something of his working method and his division of the sciences.
After specifying the subject matter, he used the positions of his
predecessors as a basis for a dialectical examination of opposing
principles. His *Physics,* for example, begins with the problem of
change and steers a middle course between Parmenides, who argues
that there is only one principle of nature and no real change, and
the atomists, who argue for an unlimited number of principles.
Aristotle settled for three principles: matter, form, and privation by
arguments that sometimes analyze change as observed and sometimes analyze the language used to describe change. In other cases
his search for principles often terminated in hypotheses, some of
which were simply bad guesses. Thus he explained all astronomical
phenomena in terms of an intricate series of interacting geocentric
spheres, while he explained all meteorological phenomena in terms
of two principles, moist and dry exhalations produced by the earth.
He did not perform experiments and would undoubtedly have considered it unnatural to study bodies and activities under laboratory
conditions rather than in their natural environment; nor did he rely
on mathematical analyses.

In the biological sciences, Aristotle's methodology proved more
successful. He believed that all sublunar beings were intrinsically
graded in a hierarchical order according to the perfection of their
forms. The four elements, earth, water, air, and fire, were at the
base, while man, a rational animal, was at the apex of the hierarchy.
The intermediate forms could be known scientifically only by careful
classification into species. To this end Aristotle performed careful
dissections and classified some five hundred and forty animal species, supplying such a wealth of precise information that Darwin
considered him the greatest of all naturalists. Such studies strongly
reinforced his view that the universe was an ordered cosmos best
understood by analogy with a living organism.

In this way Aristotle fused (1) a methodology that was suffi-
ciently open to allow further development and even some conceptual
revision; (2) a deductive structure that, at least in principle, guaranteed a realistic interpretation of science in terms of causes; and (3)
a world view that integrated the different sciences into a coherent

hierarchical order. Aristotle formed the first physical synthesis, a synthesis that encountered no serious competitor for two thousand years. We shall consider the events of this interval only in terms of their subsequent bearing on the Aristotelian interpretation of science.

The Fate of Aristotelianism

The Museum, which was founded in Alexandria by Alexander the Great's general, Ptolemaios, was the first government-sponsored institute for scientific research. Its early directors, Demetrios and Straton, were trained in Aristotle's Lyceum, and therefore Aristotle's influence was strong in the new scientific capitol of the world, though his scientific and philosophical lectures were not published for over two hundred years. Aristotle's teaching on the structure of scientific systems seemed to find its perfect realization in the deductive systems of Euclid and Archimedes. However, the directions that Aristotle had given for the proper construction and interpretation of scientific systems did not describe the practices actually employed by these scientists of the Hellenistic renaissance.

The works of Archimedes of Syracuse, the greatest physicist of antiquity, provide an interesting test case for the Aristotelian methodology. In a little treatise discovered only in the present century, Archimedes described the method he used in making the scientific discoveries on which his axioms were based. He called it a 'mechanical method'. In contemporary terms, he relied on imaginative models. Thus he measured the areas of plane figures by weighing their shapes cut out of uniform material. The reliance on models was an aid to scientific discovery, but was not probative. Archimedes was clear on this point.

> This procedure is, I am persuaded, no less useful even for the proof of the theorems themselves; for certain things first became clear to me by a mechanical method, although they had to be demonstrated by geometry afterwards because their investigation by the said method did not furnish an actual demonstration.[3]

3. Thomas L. Heath, ed. and trans., *The Method of Archimedes*, orig. trans. Heiberg (Cambridge: Cambridge Univ. Press, 1913), p. 13.

An examination of the axioms presented by Archimedes in his physical treatises, "On the Equilibrium of Planes" and "On Floating Bodies," reveals that the axioms actually refer, not to reality itself, but to idealized models of reality, such as a perfectly spherical earth composed of water, or an ideally uniform plane balanced on a perfect knife-edge.[4]

Realism, as such, does not seem to have been a problematic issue for Archimedes. Yet when reflected upon, his practice tended to undercut a key aspect of Aristotelian scientific realism, the insistence that axioms expressing causal relations are the basis of explanation. Archimedes' axioms were not attained by, or interpreted in, terms of an Aristotelian type of causal analysis. They were merely postulated and justified by their intrinsic intelligibility and by the conclusions following from them.

Still less did developments in astronomy accord with the Aristotelian charter. The Alexandrians developed or improved instruments, such as the astrolabe and the diopter, that made precise astronomical observation possible. These observations, coupled with the records of earlier and surprisingly accurate Babylonian observations, convinced them that the theory of geocentric spheres did not fit the observed inequalities in planetary motions. Aristarchos of Samos (2nd century B.C.) developed a heliocentric theory that broke with the Aristotelian astronomical system but seemed to fit, at least in part, into the Aristotelian pattern of explanation. He tried to base mathematical astronomy on what he thought was a physically true explanation of the universe. His successors did not take this heliocentric view seriously, partially because any doctrine of perfect circles could not fit the observed irregularities in planetary speed and position, and partially because the geocentric view was "obviously" true. The great astronomers of late antiquity, Apollonios, Hipparchos, and Ptolemy, effectively made astronomy a branch of pure mathematics (in the ancient sense of the term), rather than a physically true system of the world. Ptolemy, who had access to Aristotle's printed works, was explicit on this point. As a foundation he accepted the manifest phenomena (including the immobility

4. Thomas L. Heath, ed., *The Works of Archimedes* (Cambridge: Cambridge Univ. Press, 1897). Archimedes' axioms for "On the Equilibrium of Planes" are on pp. 189–90, and those for "On Floating Bodies" are on pp. 253–300.

and sphericity of the earth) and the accumulated data, and he attempted to find a geometry that saved the phenomena rather than give a realistic account.[5]

What bearing do these developments have on Aristotle's synthesis in general and on his doctrine of scientific realism in particular? This question was a central concern of the Aristotelian commentators of late antiquity, Iamblichus, Simplicius, John Philoponos, and others. Whereas Aristotle had assigned mathematics a supplementary role in natural science, these critics gave it a primary role. Simplicius defended an interpretation of science that resembles, and perhaps influenced, contemporary hypothetical-deductive explanatory views.[6] One reason these critics gave for their modifications of Aristotelian realism (or which they attributed to Pythagoras and Plato on whom they projected their views) was the desire to make theories of scientific explanation consistent with the methods that had proved successful.

Early scholastic philosophy, dominated by St. Augustine's adaptation of Platonism, sought the intelligibility of nature in something that transcended observable reality. For the English philosophers, Grosseteste, Kilwardby, and Roger Bacon, this meant a platonic physics that stressed mathematical, rather than physical, reasoning. However, the rediscovery of Aristotle's writings transmitted by the Arabic philosophers soon established Aristotelianism as the basis for the recovery, development, and interpretation of science.

In the writings of St. Albert the Great, St. Thomas Aquinas, and the fourteenth-century nominalists, there emerged a transformed Aristotelianism at once more spiritualistic and more materialistic than the original formulation. The new spiritualism stemmed from the doctrine of a transcendent God, Creator of the universe, in contrast to Aristotle's First Mover, who was essentially a part of the universe. This doctrine, rather surprisingly, led to materialism. For

5. In the preface to his *Almagest* (*Encyclopaedia Britannica*, Great Books, vol. 16, p. 5), Ptolemy accepts the Aristotelian distinctions between physics, mathematics, and theology (or metaphysics) and places his work on the level of mathematics. This clearly means that he did not intend to give a physical explanation.

6. A summary of these views may be found in S. Sambursky, *The Physical World of Late Antiquity* (London: Routledge & Kegan Paul, 1964). Contemporary treatments of science as a hypothetical-deductive system stem from P. Duhem, whose historical writings made Simplicius known in the twentieth century.

Plato and Aristotle, matter in itself was unintelligible, whereas the doctrine of creation implied that whatever was created, even matter, must be intelligible. This faith in an intrinsically intelligible universe launched a search for natural laws and led to a de-emphasis, though not a disappearance, of organismic explanations.

The redevelopment of physics gradually transformed and eventually shattered this matrix of medieval Aristotelianism. We shall only consider those aspects of the development that bear on the central tenet of Aristotelian or scholastic scientific realism, the doctrine that true scientific explanation must be based on causal knowledge. The focal point of criticism in the fourteenth and fifteenth centuries was the Aristotelian doctrine of motion, both terrestrial and celestial. Aristotle had divided terrestrial motion into natural motion, such as bodies tending toward their natural place, and violent motion, such as throwing a spear. His theory of antiperistasis provided a causal explanation of violent motion. The spear-thrower transmits a moving force to the medium so that the air, which parts to let the spear pass and then gives the spear a push as it closes in behind it, is the continuing cause of the motion.

Criticisms of this explanation culminated in the 'impetus' theory of Jean Buridan, where the cause of violent motion was sought in the impetus given the projectile rather than in the medium. Galileo eventually inherited this tradition, but explained motion in terms of inertia (or momentum), rather than impetus. This had a two-fold significance. First, whereas impetus was conceived of as a cause of motion, inertia was a mathematical specification rather than a causal force. Second, Galileo was successful; his law worked. This success, coupled with his popular and usually polemical explanations, contributed in no small measure to the change from a predominantly qualitative physics, based on real or reputed causes, to a quantitative physics, based on experimentation and mathematical analysis.

In astronomy, Aristotelianism suffered a similar reversal. Some scholastics, like St. Thomas, tried to keep both Aristotelianism as a causal explanation of celestial motions, and the Ptolomaic calculations as a means of saving the phenomena. This unstable fusion faded long before Copernicus developed his heliocentric system. Copernicus himself gave no causal explanation of planetary motion. Kepler attempted to fill this gap with a bewildering variety of hypotheses: some mystical; some mathematical, in accord with Plato's idealization of mathematics; and some causal, in terms of a theory of magnetic forces. While these failed, his mathematical analysis of the data on Mars's orbit led to Kepler's three laws of motion. Here again, the search for mathematical laws, rather than the search for causal explanations, proved to be the key to progress.

Newton and the Second Physical Synthesis

Aristotelianism, the first great physical synthesis, was replaced by Newtonianism, which represented not merely a new physics but also a novel synthesis of the conceptual elements we have been considering. The tradition of scientific realism, associated with this Newtonian physics, can be adequately understood only in the light of this complex of ideas. As a matter of convenience we shall treat these ideas under the headings of methodology, content, structure, and world view.

Newton's infrequent discussions of scientific methodology, some of which are reproduced here, were influenced by Francis Bacon's stress on empirical methods and by the tradition of mathematical analysis stemming from Galileo, Kepler, and Descartes. More influential, however, was Newton's reflection on his own experiences, their successes and failures. Two investigations played a crucial role in the development of his methodology. The first was his analysis of light. By a 'crucial experiment' (a term Newton adopted from Bacon) of passing a pencil of white light through two separated prisms, the first of which refracted white light into a spectrum of colors, while the second bent selected rays of different colors but introduced no new colors, Newton thought he had proved that white light is really composed of separate colors with different indices of refraction. When his critics labelled this a hypothesis, Newton objected vehemently, insisting that it was not a hypothesis but a physical conclusion proved by induction from phenomenon. He seems to have convinced only one of his critics, Ignatius Pardies, and this negative reaction contributed to his extreme sensitivity in the use of the term 'hypotheses'.

Newton's protracted analysis of gravitation seems to have been even more decisive in shaping his interpretation of scientific methodology. Gravitation, one of the day's leading scientific problems, was being tackled along two distinct lines. The first was essentially physical and hypothetical, an attempt to explain gravity by postulating a causal explanation of gravitational attraction. Descartes led this movement with his hypothesis of vortices in space moving the planets. The second approach was essentially mathematical. In England, Robert Hooke, Edmund Halley, and the architect, Christopher Wren, met regularly to discuss such problems and by 1684 decided that the force by which the sun kept the planets in their orbits should obey an inverse law.

Independently and privately Newton was trying both approaches. To explain the cause of gravity, he first introduced the hypothesis of an ether whose increasing density, proportional to the distance from the sun, produced a net force inward. He abandoned this because it entailed untenable conclusions. His second causal hypothesis, based on active powers of bodies, proved even less acceptable. At the same time he developed the inverse square law and showed that it gave approximately correct answers, provided that the distances were large enough so that the bodies could be represented by point sources, a rather crude approximation for planetary systems. After developing the calculus, he proved, to his own surprise, that his inverse square law was exact rather than approximate and that he could, as Halley had requested, derive Kepler's law of elliptical motion from the assumption of an inverse square law.[7]

The result was that he had a rigorous law, apparently of universal validity, and yet had no notion of the cause of gravity. Thus the most successful scientific treatment of this basic problem violated a fundamental requirement of scientific explanation, held by interpreters from Aristotle through Bacon, that science must be an explanation in terms of causes. This could have led to the abandonment of the requirement. Historically, however, it contributed to the transformation of the meaning of 'causality' from the Aristotelian-scholastic notion of efficient causality as activity by an agent, to the mechanistic notion of causality as invariable succession in accord with rules, a notion that Hume later criticized.

The methodology developed by Newton can be summarized in three steps: *experimentation*, by which the pertinent phenomenon

7. This interpretation of Newton's development is based on Melbourne G. Evans, "Newton and the Cause of Gravity," *The American Journal of Physics,* 26 (1958): 619–24.

is isolated and measured; *induction* of general laws from phenomena; and *composition,* or a deductive explanation of the observed phenomenon from general laws. Hypotheses have no place in physics. But here Newton uses 'hypotheses' in a restricted sense referring especially to postulated causal mechanisms, such as the various theories of the cause of gravity, which were not experimentally established in the way he established the composition of white light.

The content of Newton's physics, specifically of his mechanics, was a set of mathematical laws and conclusions that explained and predicted the behavior of material bodies subjected to certain types of forces. Although this now seems commonplace, it was startling when viewed in historical perspective. In scholastic physics (or philosophy of nature), properties and characteristic activities of bodies were to be explained in terms of their specific natures. Natures, so conceived, play no role in Newtonian physics, where all bodies were treated as if they were nothing but a collection of material particles with such primary qualities as mass, extension, gravitational attraction, and some sort of cohesive force. As the queries appended to his *Opticks,* as well as some of his other writings show, Newton clearly realized the limited nature of these simplifying assumptions. The success of his method, however, provided a constant temptation to drop the qualifying "as if" and conclude that all bodies are nothing but ordered aggregates of particles (with the option of some form of dualism to explain man's spiritual qualities).

The structure of Newton's *Mechanics* might seem to represent the ideal fulfillment of the program charted in the *Posterior Analytics,* running from clearly stated axioms, notably the three laws of motion and the law of gravity, to conclusions derived by rigorous deductive methods. The interpretation of this structure, however, was only semi-Aristotelian. Newton's axioms were, as the title indicated, mathematical principles of natural philosophy, a very un-Aristotelian notion. Such principles were to be sought by experimentation and mathematical analysis, rather than by dialectics and causal analysis. Yet one Aristotelian doctrine remained, the requirement that the principles must be true, primary, and certain; and this was the guarantee of the truth and certainty of the conclusions.

As Newton himself seems to have realized, at least to some degree, his idea of 'inducing' or 'deducing' general laws from phenomena was radically incapable of justifying the conclusion that his laws of motion were universally valid laws governing the motions of all material bodies. Historically, however, this issue was not decided through a critical analysis of the methodology by which these laws were attained or corroborated. They were, rather, interpreted by and justified through a new *world view:* a mechanical universe, a vast clockwork system, ultimately composed of mass particles whose activities and interactions obeyed mathematical laws. This was the vision of an age, and Newton crowned and completed that vision by laying bare the basic laws that ruled the world machine.

Such, in simple outline, was the physical realism of classical or pre-twentieth-century physics. The basic laws of science, Newton's laws in particular, were considered true, primary, and certain because they were laws of nature built into the fabric of the cosmos. Discovering a law, in this context, was similar to discovering America. Both were already there waiting to be discovered. Newtonianism soon triumphed over the competing views of scientific explanation proposed by Descartes and Leibniz.

Scientific realism, which had been a theory of philosophers, became an article of faith for the practicing scientist. Euler, Lagrange, D'Alembert, Laplace, and the other eighteenth-century physicists could not extend Newton's physics without accepting the foundation on which this work was based. The keystone in this foundation was the doctrine that the laws of physics were laws of nature's, not of man's, designing. That these laws enabled one to predict accurately the future positions of planets forcibly demonstrated the validity of this interpretation. As Laplace, the greatest of the Newtonians, summed it up:

> It is from a comparison of facts with each other, by attentively considering their relations, and by this means reascending to phenomena which are continually more and more extensive, that at length we have been able to discover these laws which are continuously impressed on the various effects they produce. Then it is, that nature by revealing herself, shows how the infinite variety of phenomena which have been observed, may be traced up to a small number of causes, and thus enables us to determine antecedently those effects which ought to be produced; and being assured that nothing will derange the connection between causes and their effects we can extend our thoughts forward to the future, and the

series of events which shall be developed in the course of time will be presented to our view. It is surely in the theory of the system of the world that the human mind has, by a long train of successful efforts, attained to this eminence.[8]

Though physicists generally accepted the Newtonian paradigm of scientific explanation and the world view it entailed, philosophers tended to be less tractable. Leibniz, Berkeley, and Hume criticized Newton's views on absolute space and time. Immanuel Kant, who fully accepted Newtonian physics except for the idea of deducing physical conclusions from mathematical principles, clearly realized that induction from phenomena could not justify the necessity and universality attributed to the laws of physics. In his *Critique of Pure Reason* he labelled such laws 'synthetic *a priori*' propositions and sought the ground of the possibility of such principles in the propensity of the mind to order experience by the imposition of forms of sensibility, categories of understanding, and rules of reason. While his penetrating criticism undercut the simple realism accepted by classical physicists, it had little influence on scientists until much later.[9]

The nineteenth century witnessed the first flowering of chemistry and the development of new physical sciences, notably thermodynamics and electromagnetic theory. Though some new discoveries, such as the wave theory of light, contradicted particular details of the Newtonian view, these new developments could, it was believed, support mechanistic realism if the new sciences could be reduced to Newtonian mechanical principles. Such a program won a partial success in thermodynamics. Joule, Mayer, and Helmholtz proved that heat is a form of energy. On this foundation, Maxwell, Gibbs, and Boltzmann explained the gas laws of Boyle, Charles, Gay-

8. P. S. de la Place, *The System of the World*, trans, H. H. Harte, vol. 1 (London: Longmans, Green, 1830), p. 72.

9. Kant's ideas on natural science presented in his *Preface to the Metaphysical Foundations of Natural Science* were more influential than his *Critique* in the development of nineteenth-century science. For a summary of these ideas see Joseph J. Kockelmans, ed., *Philosophy of Science: The Historical Background* (Glencoe, Ill.: Free Press, 1968), pp. 9–28. Their historical influence is traced by Joseph Agassi, *Towards a Historiography of Science* (The Hague: Mouton, 1963). Toward the end of the nineteenth century H. Hertz endeavored to develop mechanics along the *a priori* lines specified in the *Critique*. See his *The Principles of Mechanics* (New York: Dover, 1965).

Lussac, and others by the assumption that a gas is a collection of molecules whose motions and interactions are strictly governed by the laws of mechanics. Only electromagnetic theory resisted reduction to mechanics. This, however, was looked on as a temporary difficulty, rather than a defect, in the foundation. As soon as the existence of ether was conclusively established, it should be possible to give a mechanical, i.e., a realistic, explanation of electricity, magnetism, and light. Optimism reigned. Yet doubts and difficulties were beginning to appear. In the subsequent section we shall consider how their growth led to the dissolution of the second physical synthesis.

Breakdown of Mechanism

Both the systematic part of this introduction and the readings that follow focus on contemporary problems. This historical outline, accordingly, will omit detailed explanations of any matter to be treated subsequently. For convenience, this section will separate philosophical currents from scientific developments and consider each only from the viewpoint of its bearing on the problem of scientific realism.

Auguste Comte (1798–1857) introduced the doctrine of positivism in terms of the three stages of human intellectual development: *theological*, in which explanations of natural phenomena are based on supernatural forces; *metaphysical*, in which nature is considered a regulative system of general forces known by general concepts; and *scientific*, in which the concentration is on facts and their interrelation by mathematical laws. John Stuart Mill (1806–1873) took up Comte's emphasis on facts, but gave them a psychological interpretation that, he felt, would serve as a basis for laws of induction.

Around the turn of the century three scientists-turned-philosophers developed more penetrating criticisms of classical realism. Ernst Mach (1836–1916) elaborated the positivist stress on facts, which he interpreted as nothing but sensations. Ideas, including scientific laws and theories, were explained by his 'principle of economy' as simply a useful summary of actual and possible experiences. At the same time the French mathematician-philosopher, Jules

Henri Poincaré (1854–1912), was developing a doctrine of conventionalism. The subject matter of science, as he interpreted it, is neither things nor sensations, but relations between things. These are knowable by intuition; things in themselves are not. Such intuitions in physics lead to empirical generalizations that are approximately true of individual cases. When successful, these generalizations are promoted into 'conventions,' or disguised definitions, of key terms such as 'point' and 'line' in geometry, or 'mass' and 'force' in mechanics. Such conventional laws are not falsifiable, for they are no longer empirical statements in the strict sense.

Pierre Duhem (1861–1916) distinguished between a representative part and an explanatory part in physical theory. Only the former, the formal mathematical structure, is truly a part of physics. The latter is really a disguised and dispensable intrusion of metaphysics. His interpretation of scientific theories as formal hypothetical-deductive systems and his stress on the problems of verification and falsification have had a strong influence in recent years.

Although these critics disagree on many details, they basically agree on one fundamental point. The intelligibility of a scientific theory cannot be considered a reflection or a transcription of an intelligibility intrinsic to material reality. Rather, it is supplied by the scientist himself in constructing a deductive system that yields experimental laws and observational results as conclusions drawn from general principles.

In both the Aristotelian and the Newtonian syntheses, a doctrine of scientific realism was based on a 'top-of-the-page' interpretation of scientific systems. Though Aristotelians explained basic laws causally and Newtonians mechanistically, both insisted that the fundamental laws of a scientific system must be true, primary, and certain. They had these qualities because they were laws of nature, and their truth, in turn, guaranteed the truth of the conclusions flowing from them. The new critics repudiated this shared ground of scientific realism. Whether one considered the basic principles of a physical science to be an economical summary of actual and possible facts (Mach), conventions (Poincaré), or the axioms of a hypothetical-deductive system (Duhem), it was not meaningful to speak of them as 'true'. Rather than explaining and justifying the conclusions drawn from them, the principles were themselves justified, in the sense of being considered more or less adequate, by the correspondence between the conclusions they yielded and experimental observations. New developments in physics, which Mach, Duhem, and, for a period, Poincaré opposed, temporarily eclipsed the cogency of these criticisms. Yet they still present a challenge that any doctrine of scientific realism must meet.

At the time these criticisms were being developed, mathematics was undergoing a conceptual revolution with the development of non-euclidean geometry by Lobachevsky, Boylai, and Reimann, the formal redevelopment of euclidean geometry by Hilbert, Cantor's theory of transfinite numbers, and Peano's axiomatization of arithmetic. From this ferment emerged new views on the nature of formal systems: logicism, formalism, and intuitionism, views which would eventually condition the twentieth-century philosophy of science. Logicism, as an interpretation of mathematics, takes its name from the 'logistic thesis' introduced by Frege in 1884 and independently by Bertrand Russell in 1902. The thesis contends that the truths of mathematics can be derived from logic alone, a contention that was heroically, though somewhat unsuccessfully, implemented in the monumental *Principia Mathematica* of Whitehead and Russell (1910–1913). The ideal held aloft by this thesis of a formal unification of scientific knowledge by logical reduction to primitive elements set a pattern for much future philosophical thinking.

The logical atomism of Bertrand Russell was the first significant attempt to translate this ideal into a full-fledged philosophy. In accord with the logistic method, he attempted to reduce the truth value of all complex sentences to the truth values given by atomic sentences and logical connectives. The truth value of these atomic sentences is decided on such non-logical grounds as comparison with direct observation. Thus, the truth value of any theory is a function of the correspondence between its irreducible atomic sentences and certain observed or postulated features of the world.[10]

Russell's development of logical atomism, unfortunately, was spotty and inconsistent. A rigorous following of his theory of meaning would deny meaningfulness to the sentences in which the theory was expressed. More rigorous, more consistent, and much more diffi-

---

10. See B. Russell, "Logical Atomism" in A. J. Ayer, *Logical Positivism* (Glencoe, Ill.: Free Press, 1959), pp. 31–50. A detailed criticism may be found in J. O. Urmson, *Philosophical Analysis: Its Development between the Two World Wars* (Oxford: Clarendon Press, 1956), pp. 1–98.

cult to interpret, was the logical atomism presented in Ludwig Witt-genstein's *Tractatus Logico-Philosophicus*.[11] We will not attempt to explain his views here, but merely indicate the fate of scientific realism in his system. The whole of natural science is the totality of true propositions. Such propositions are ultimately reducible to elementary propositions, which are concatenations of names ex-pressing a state of affairs. The world is a collection of facts, of which a logically perspicuous language is a mirror image. Scientific laws and theories have only an indirect relation to reality in specifying the pattern of propositions that give the simplest description of reality. Philosophy is not a natural science, does not have a body of doctrine, and gives no information about the world; it is simply the activity of clarifying the meaning of sentences.

Logical positivism, the school of thought that until recently domi-nated the field of the philosophy of science, represented a fusion of the empiricist tradition stemming from Hume, Mill, and Mach, with the new logic as interpreted by Wittgenstein. From the *Trac-tatus*, the logical positivists, M. Schlick, R. Carnap, O. Neurath, and H. Hahn, adopted the distinction of all meaningful propositions into those that are analytic and those that are empirical or synthetic. Analytic propositions are either tautologies or contradictions and give no factual information. Empirical propositions are either re-ports of sense experiences (protocol sentences) or finite generaliza-tions from such reports whose truth value can be determined only by experimental verification. Since metaphysical and theological statements fit into neither category, they were dismissed as mean-ingless pseudo-propositions. Because a robust doctrine of scientific realism cannot even be formulated in sentences that accord with this school's theory of meaningfulness, realism was dismissed as a pseudo-problem.

Logical positivism soon encountered serious difficulties and ob-jections that engendered a series of modifications. Yet, simplistic as it was, the positivistic platform was quite influential. Here, for the first time, was a growing group of competent men making the phi-losophy of science their central concern, agreeing on a basic pro-gram and set of doctrines, and promulgating their new creed through journals and international meetings with an apostolic zeal. Contem-

11. Ludwig Wittgenstein, *Tractatus Logico-Philosophicus*, trans. D. F. Pears and B. F. McGuinness (London: Routledge & Kegan Paul, 1961).

porary defenses of scientific realism tend to be, in large part, reactions to the positivistic crusade against any species of metaphysics. However, before considering contemporary views, we must see something of the scientific developments that conditioned the present problem of scientific realism.

Relativity and quantum mechanics are the overarching contributions of twentieth-century physics. They will be considered here only with respect to their bearing on the problem of scientific realism. The special theory of relativity, first presented in 1906, is based on two postulates. The first is the invariance of the speed of light. This means that any observer moving at constant velocity would get the same result for the measured speed of light regardless of how fast he was moving either toward, or away from, the source. This postulate is certainly not in accord with simple common-sense realism. The second postulate is the covariance of physical laws with respect to inertial transformations. This means that all valid physical laws must have the same mathematical form regardless of which constant velocity space-time framework is chosen.

This theory radically clashed with some assumptions basic to the Newtonian, or mechanistic, world view. Gone was absolute space and time; gone too was the ether, the putative foundation for electromagnetic vibrations. Yet these notions had played an essential role in classical physical realism, particularly in supporting the idea that the equations of physics gave an objective description of physical motions as they really are.

The general theory of relativity that Einstein developed in 1916 was too remote from ordinary concerns to have an immediate impact. Its speculative nature, however, strengthened Einstein's growing distrust of positivistic interpretations of science. In his early years, Einstein had been strongly influenced by Mach's criticism of Newton. In his later years, he tended to stress the creative role of the scientist in forming new hypotheses. That such creations of reason correspond to reality was, Einstein thought, something of a miracle. Yet on the miracle that the world can be comprehended by reason rests the possibility of science. The selection included here contains one of the most concise presentations of his later views.

The quantum theory was even more revolutionary in its implications. The theory of relativity had demonstrated that the basic laws of Newtonian mechanics were approximations valid for bodies moving at speeds which are slow compared to the speed of light. Quantum mechanics demonstrated that these same laws had an approximate validity only for physical objects large enough so that they were not significantly disturbed by the process of measurement. Together, the two theories undercut the scientific basis of the second physical synthesis, the presumed universal validity of Newton's laws.

The negative aspect of this conceptual revolution was not, in itself, enough to undercut faith in a simple scientific realism. One might still believe that the basic laws of physics were mathematical transcriptions of natural laws built into the fabric of material reality, in much the same way that an architect's plan specifies the structure of a building. The true believer might simply substitute the new laws for the old while preserving the method of interpreting physical laws that grounded and justified classical scientific realism.

In spite of persevering efforts to do so, quantum mechanics could not be accommodated to such a simple objectivist interpretation of scientific laws. The article by MacKinnon traces the historical development of the controversies surrounding the interpretations derived from, or read into, quantum theory. Here was a novel challenge for any doctrine of scientific realism. Quantum mechanics treated physical reality at a more fundamental level than any theory that preceded it, and it did so with a striking and ever-growing success. Yet its discoverers and developers insisted that quantum theory did not give an objective picture of reality and that, in fact, no picture, model, or physical description of atoms and sub-atomic particles could be both correct and consistent.

A simple realism based on a mechanistic interpretation of physical laws does not fit modern physics. But this conclusion, which few will deny, need not constitute a refutation of realism. It can also be accepted as a challenge, a stimulus to develop a critical, rather than a naive, realism. Bunge and Heelan, both philosophers with a professional competence in quantum physics, have attempted to do this. Each in his own way defends a critical realism developed to answer the peculiar difficulties presented by quantum theory. The summary accounts included in this anthology represent the conclusions each has reached in a series of longer and more technical studies.

Contemporary Philosophical Positions

In recent years the philosophy of science has emerged, particularly in the United States, as a flourishing discipline characterized by a spectrum of positions and schools. Here we shall merely indicate some of the more influential contemporary opinions on the problem of scientific realism.

It was the logical positivists, more than any other group, who stressed the importance of the philosophy of science and stimulated further development. They specified the problems to be treated and, at least until very recently, tended to dominate philosophical public opinion on these problems. Because of internal development and external criticism, the original, rather simplistic program of logical positivism underwent a series of modifications. Under close analysis their sharp dichotomies between analytic and synthetic, necessary and contingent, theoretical and observational, and *a priori* and *a posteriori* revealed fuzzy borders. Later modifications even seemed to call in question the most distinctive and polemical doctrine of positivism, the contention that metaphysics is meaningless and that realism, in particular, is a pseudo-problem. Two complementary aspects of this problem may be brought out by an exposition of two authors represented here.

Carl Hempel has shown an abiding concern with the problem of the truth value of scientific statements and the criteria according to which this truth value is judged. Partially through the stimulus of Hempel's critical probing, especially at the later stages, the criterion of meaningfulness for scientific statements changed first from strict verification to verification in principle, then to confirmability, and finally to inclusion in an empiricist language. The basic idea behind the latest requirement is that if a language could be constructed in accord with the empiricists' criteria of meaningfulness—something that has not been done—then any sentence expressed in this language would automatically be meaningful. Though few believed that scientific principles could be expressed in such a language, many hoped that scientific conclusions could be. Concern with such questions led Hempel to search for the basic conditions that any theory must fulfill to be both true and empirically meaningful, a contemporary analogue to what Aristotle attempted in his *Posterior*

Analytics. The selection included here contains his simplest presentation of these requirements.

One of Carnap's continuing interests was the problem of the unit of meaning. His early stress on such individual units as phenomena or protocol sentences soon passed, and systems, rather than statements, became his focus of concern. In his *Logische Syntax der Sprache* he presented as his goal the replacement of traditional philosophy by the logic of science, interpreting the logic of science as nothing but the logical syntax of the language of science.[12] Later he supplemented *syntax,* a study of the formal rules governing the language of science, with *semantics,* which, in his interpretation, treats the correspondence rules relating linguistic systems and extra-linguistic entities. Later still he added *pragmatics,* which includes the users of language and treats such questions as use, belief, and acceptance. This conceptual expansion presented a difficulty. It might seem that scientific realism reemerged as a real, rather than a pseudo, problem with the admission of extra-linguistic entities to which one who accepts the theory is logically committed.

In the article included here, Carnap attempts to dispel such fears. Although his conclusions have not been widely accepted, his manner of treating the problem has become almost normative. With respect to a particular system, e.g., atomic physics, he distinguishes between *internal* questions formulated within the system and *external* questions asked about the system. "Do atoms really exist?" considered as an internal question can admit only an affirmative answer. But this is analytic rather than factual. Since atoms are an ontic commitment of the theory (their names serve as values for the bound variables in existential axioms), any denial of their existence would be self-contradictory within this system. If the same question is viewed as an external question, then it is incorrectly phrased. The operative question is, "Should the system be accepted as explanatory?" This calls, not for a metaphysical commitment, but for a practical decision. Neither answer, accordingly, affords a foothold to any doctrine of metaphysical realism.

In spite of the changes and transformations that have occurred, logical positivism remains at a bottom-of-the-page interpretation of

12. The English edition, *The Logical Syntax of Language,* trans. Amathe Smeaton (London: Routledge & Kegan Paul, 1937) is an expanded version of the published German edition.

scientific systems. This is most explicit in Carnap,[13] who views a scientific system as a formal calculus to which he gives a physical interpretation by means of rules that establish a correspondence between conclusions derived within the theory and observation statements. In recent years this manner of interpreting scientific systems has been severely attacked on various grounds. First, it relies on a sharp distinction between analytic and synthetic statements. Quine's outright denial of this distinction has triggered a massive re-examination of the whole question. Secondly, it effectively denies scientific theories a truly explanatory function. In this view, an abstract theory has no physical significance until it is translated into observational terms by means of correspondence rules. The observational framework, accordingly, contains all the information there is, while a theory is essentially an inference mechanism leading from one set of observation statements to another.[14]

These are the relatively superficial difficulties. More disturbing are the presuppositions and implications of this correspondence approach with respect to both frameworks. If terms within the theoretical framework acquire physical significance by virtue of correspondence rules, what of such terms as those naming theoretical entities that have no observational language correlates? They can only be considered theoretical constructs or useful fictions. Yet scientists generally do believe in the existence of atoms, although they reject visualizable models. Secondly, since the observational language serves as the basis for interpreting any theories, it is logically shielded from any real revision. Whence this absoluteness? In Sellars's diagnosis, it stems from the myth of the given, the epistemologically naive idea that the category terms used to make observation reports are somehow given in immediate experience.[15] The

13. See R. Carnap, "The Methodological Character of Theoretical Concepts" in H. Feigl and M. Scriven, ed., *Minnesota Studies in the Philosophy of Science,* 1 (1956): 38–76.

14. These difficulties are discussed from a logical empiricist point of view by Carl Mempel in "The Theoretician's Dilemma," *Minnesota Studies in the Philosophy of Science,* 2 (1958): 37–98.

15. In addition to the article included here, Sellars treats this difficulty in detail in "Empiricism and the Philosophy of Mind" in his *Science, Perception and Reality* (New York: Humanities Press, 1963), pp. 127–96. Current debate on the observation-theory dichotomy is summarized in Dudley Shapere's introduction to his *Philosophical Problems of Natural Science* (New York: Macmillan, 1965), pp. 1–29.

article by E. E. Harris treats these epistemological problems from the point of view of an integrated, coherent theory of knowledge.

The first step away from a bottom-of-the-page interpretation is the middle-of-the-page interpretation developed in different ways by Campbell, Braithwaite, and Nagel. Nagel develops this by stressing the differences between empirical laws and theories. Empirical generalizations, such as Boyle's law, Kepler's laws, etc., can be expressed in one statement, can be established in a quasi-inductive way, can be verified or falsified in a relatively direct and unambiguous fashion, can precede the theories devised to explain them, and can survive their demise. A theory, on the other hand, such as the kinetic theory of gases, has a complex logical structure, admits of no simple inductive justification, and is confirmed or refuted by experiments only in a partial, and rather indirect, way.[16]

This dichotomy gives a basis for interpreting theories by a middle-of-the-page correspondence. From the axioms of a theory, such as the kinetic theory of gases, one can deduce general conclusions that have the same form as independently established empirical generalizations, e.g., the gas laws. Correspondence rules, such as the relation between the mean kinetic energy of molecules and the observed temperature of a gas, lead to a partial identification. For the theory to be acceptable, however, it must do more than yield the observed laws. It should also, as does the kinetic theory, correct their limitations, provide more accurate laws, and extend them to further domains.

In this view of scientific theories, as the selection from Nagel indicates, the difference between a realistic and an operational interpretation of scientific theories is not particularly significant. By appropriate definitions of such terms as 'real', one can practically dissolve the differences between the two. Sellars, in the article included here,[17] strongly objects to what he dubs the 'levels interpretation' of scientific theories, chiefly on the grounds that it confuses various senses of 'meaning'. Theories, in Sellars's account, do not explain empirical laws by deductive subsumption. Rather, they

16. See E. Nagel, *The Structure of Science* (New York: Harcourt, Brace & World, 1961), chap. 5.

17. W. Sellars gives a detailed criticism of Nagel's position in his "Scientific Realism or Irenic Instrumentalism," *Boston Studies in the Philosophy of Science,* vol. 2, ed. R. S. Cohen and M. W. Wartofsky (1965): 171–204.

explain why observable things obey empirical laws to the extent they do. Thus, a gas obeys Boyle's law because, as kinetic theory teaches, it is really a collection of molecules in random motion. Acceptance of this interpretation implies acceptance of the objects postulated or presupposed by the scientific framework.

Sellars's position on this problem represents one of the pivotal points in a new and complex philosophical synthesis that he is gradually developing.[18] The sense in which it constitutes a neorealism can be seen by contrasting his position with Carnap's. Like Carnap, Sellars holds that knowledge, or at least the public expression of knowledge, can best be analyzed in terms of an interrelated collection of linguistic systems, ranging from ordinary language to the physics of fundamental particles. But instead of simply accepting ordinary language, Sellars submits it to a critical logical and epistemological investigation, extracts the view of the world implicit in this language, and considers the various philosophical systems directly or indirectly based on this view as competitors with the scientific world view. Here Carnap's internal-external distinction becomes operative, though its significance is revised. While accepting a theory as explanatory is a pragmatic decision, it should also be a reasonable decision. One who has good reasons for accepting a theory as basic also has good reasons for accepting the entities postulated by the theory. Accepting fundamental scientific theories as ultimate explainers, accordingly, entails a realism, an acceptance of the entities postulated by these theories as existentially real.

Quine's views are somewhat similar in that he accords ordinary language a methodological, but not an interpretative, primacy. Because of Quine's rejection of the analytic-synthetic distinction, he does not admit of a sharp division of knowledge into separate systems. There is, rather, the great sphere of knowledge. Both in developing and interpreting this, Quine relies strongly on the prag-

18. An outline of Sellars's views may be found in Richard Bernstein, "Sellars' View of Man in the World," *The Review of Metaphysics*, 20 (1966): 115–43, 290–316. A briefer summary coupled with a criticism may be found in E. MacKinnon, S. J., "The New Materialism," in *The Heythrop Journal*, 7 (1967): 5–26. Sellars's new volume, *Science and Metaphysics: Variations on Kantian Themes* (New York: Humanities Press, 1968) contains a further extension of his doctrine. A detailed evaluation and summary may be found in my critical review in the *Philosophical Forum*, 1 (1969): 509–45.

matic role of decisions. Rather than agonize, as others might, over the validity of a transcendental deduction leading from the conceptually basic to the existentially real, Quine simply asks which decisions allow for the simplest, most consistent, and therefore most useful integration of knowledge. The issues involved in these developments, especially those raised by Sellars, will be discussed in more detail in the second half of the Introduction. It is interesting to note, however, that the ongoing reaction against the excesses and shortcomings of logical positivism is leading to a return to realism, albeit a realism developed and interpreted in rather novel fashions.

In this survey we have concentrated on the mainstream in the philosophy of science. But there are subsidiary channels and independent streams. Foremost among the former, i.e., related to but differing from the main line of development, is the critical rationalism of Karl Popper and his followers, I. Lakatos, J. Agassi, and P. Feyerabend. Popper has developed an anti-inductivist interpretation of science that explains theories as conjectures to be retained only if all efforts to refute them have so far proven unsuccessful. Although he does not reject metaphysics, the problem of philosophical realism is not one of his central concerns.

The contemporary philosophers we have been considering generally approach the problem of scientific realism through an analysis of the structure, functioning, and interpretation of scientific systems. There are, however, some rather different approaches. Both Thomists and Marxists have developed doctrines of philosophical realism that supply interpretive frameworks and criteria used in judging scientific realism. Here we shall concentrate on Thomism and, more as a matter of convenience than of precision, distinguish three treatments of the problem of scientific realism: natural philosophy, immediate realism, and critical realism.[19]

19. The best survey of neoscholastic doctrines, particularly in epistemology, from about 1830 to World War II may be found in George Van Riet, *Thomistic Epistemology: Studies Concerning the Problem of Cognition in the Contemporary Thomistic Schools* (St. Louis: B. Herder), vol. 1, trans. G. Franks (1963), vol. 2, trans. D. G. McCarthy and G. E. Herynich (1965). A summary of scholastic positions on the philosophy of nature and the philosophy of science may be found in W. A. Wallace, O.P., "Towards a Definition of the Philosophy of Science" in *Mélanges à la Mémoire de Charles De Koninck* (Quebec: Les Presses de l'Université Laval, 1968), pp. 465–85.

In spite of the historical setbacks it has encountered, a modified Aristotelian interpretation of the scientific enterprise still survives in the form of a general philosophy of nature. The relations among this branch of philosophy, the positive sciences, and metaphysics are usually explained through a doctrine of degrees of abstraction, a doctrine that derives from Aristotle's division of the sciences and the epistemological underpinning given this doctrine by the scholastic tradition. Any universal concept prescinds from the individuating features of the particulars to which it refers. The first degree of abstraction is the abstraction of a whole from its parts. A systematic development based on this level of abstraction treats these wholes in terms of their constitutive principles: natures, properties, activities, and causes. If one accepts the idea that the core concepts in this philosophy of nature are derived by immediate abstraction from experienced reality, then the philosophy of nature not only is realistic itself, but can also ground a doctrine of scientific realism. The formulation of any natural science presupposes general concepts and principles that play an essential role in specifying and describing the subject matter treated in the science, in grounding the inductive process that leads from experiments to empirical generalizations, and in supplying a general conceptual background for particular theories. Since these general concepts are part of the proper subject matter of the philosophy of nature, this philosophy should be a general science, with the natural sciences, or at least certain aspects of the natural sciences, playing a subordinate role. The precise relationship that obtains is disputed.

The second degree of abstraction is formal abstraction, or abstraction of a form from matter, and is considered the basis of mathematical reasoning. If this is granted, then a science like physics can be explained in terms of two components: a basic physics, which is essentially a conceptualization of a world of objects whose properties and interactions are determined by natures and causes; and a mathematical formulation, which should be considered more an auxiliary tool than a proper part of physics. Metaphysics was formerly explained in terms of a third degree of abstraction. Now it is generally interpreted as characterized by the distinctive role of existential judgment rather than by abstraction.

This interpretation of philosophy has encountered severe criticisms even from philosophers within the Thomistic tradition. Two rather obvious ones are the excessive reliance on Aristotelian physics and the manner of interpreting science by fitting a reinterpreted residue of scientific theories into a predetermined framework. A more fundamental objection comes from the contention that no concepts are formed by abstraction in the manner specified by the theory.[20] I believe that these criticisms are generally valid, but that this interpretation still has something to offer if redeveloped along the radically different lines considered in the second section of the Introduction. Many of the defenders of this tradition, unfortunately, seem to be more concerned with demonstrating that it is genuinely Thomistic than in coming to grips with the fundamental difficulties in their approach.

Immediate realism, best exemplified by E. Gilson and his followers, is a metaphysical approach to philosophical explanation developed to counter idealism, whether direct or disguised. As Gilson explains the position:

> The first step on the path of realism is to perceive that one has always been realistic; the second is to perceive that, whatever one does to become otherwise, one will never succeed; the third is to ascertain that those who pretend to think otherwise, think in realistic terms as soon as they forget to act their part. If one wonders why, the conversion is almost complete.[21]

The realist, Gilson continues, should steadfastly refrain from ever questioning the validity of his position. Such questions can only be

---

20. See Peter Geach, *Mental Acts: Their Content and Their Objects* (London: Routledge & Kegan Paul, 1957), pp. 11–38. In an appendix, pp. 130–31, he tries to show that St. Thomas did not hold the view of abstractionism generally attributed to him. A more detailed treatment of St. Thomas's view on this that interprets concepts as inner words produced by the mind may be found in the fourth article of B. F. Lonergan's series, "The Concept of *Verbum* in the Writings of Thomas," *Theological Studies*, 10 (1949); 3–40. W. Sellars's criticism of the 'myth of the given' in "Empiricism and the Philosophy of Mind" effectively undercuts the usual doctrine of concept abstraction.

21. E. Gilson, "Vade Mecum of a Young Realist," in R. Houde and J. Mullally, eds., *Philosophy of Knowledge: Selected Readings* (Philadelphia: J. B. Lippincott, 1960), pp. 386–94.

phrased in idealistic terms. To admit them as valid is to preclude the possibility of an acceptable answer.

A corollary of this justification is the conclusion that metaphysics must ground epistemology because being is prior to knowing. Other branches of philosophy should, in fact, be developed as applied metaphysics. This can lead to two rather different views on the problem of scientific realism. An older view, strongly influenced by Duhem, effectively combines a realistic metaphysics with a positivistic interpretation of science. Since only metaphysics really explains being, science, inasmuch as it is independent of metaphysics, should be viewed as a constructural system of knowledge based on hypotheses, or fruitful, creative fictions.

It is also possible to accept this view of philosophy and interpret science realistically. The rationale is essentially the same as for metaphysics. The scientist, like the proverbial man in the street, has a natural disposition toward realism, a disposition whose implementation actually grounds the practice and development of science. If the same scientist rejects realism while functioning as a part-time philosopher, this is chiefly because of critical difficulties that he cannot counter from his limited and specialized viewpoint. Here the metaphysician may intervene and give a more adequate grounding to scientific realism. As this argument is worked out for particular problems, it is often not very different from the natural philosophy previously considered, though it has a different internal justification. Since the realism operative in the practice of science will be considered in the next section we will postpone an evaluation of this position.

Contemporary Soviet philosophy, while it has a radically different basis, exhibits a somewhat similar defense of immediate realism. Knowledge is explained in terms of a reflection of reality in cognition, a doctrine that justifies the selective interpretation of formulations of knowledge in accord with criteria supplied by a realistic metaphysics. Like immediate realists in the scholastic tradition, these philosophers tend to refute such opponents as logical positivists by interpreting them as disguised idealists. In recent years, however, Russian philosophers have made an intensive effort to assimilate developments in Western philosophy, particularly in the philosophy of science. In the light of these assimilations, it is rather difficult for an outsider to judge whether the often polemical realism defended in the prefaces and concluding paragraphs of technical articles represents epistemological naiveté or political sophistication. It is clear, however, that Soviet philosophers are realists and are working intensively in the philosophy of science.

Critical realism, sometimes referred to as transcendental Thomism, represents a fusion of Thomistic doctrines and Kantian methods. Recent developments, building on the pioneering efforts of Scheuer and Maréchal, have assimilated much of the terminology and some of the basic ideas of Husserl and Heidegger, particularly their emphasis on subjectivity and their rejection of ontology within a naturalistic framework as critically inadequate. After a brief outline of the general methodology followed, we will indicate the relationship between this philosophy and the problem of scientific realism.

Kant's investigation of the conditions of the possibility of scientific knowledge led to a stress on transcendental aspects of knowledge. "I call every knowledge transcendental which occupies itself not so much with objects but rather with our way of knowing objects insofar as this is to be possible *a priori.*" [22] Maréchal accepted the critical question as a valid point of departure in the development of metaphysics, but sought to avoid Kant's negative conclusions about the possibility of a science of metaphysics by concentrating on the act of knowing, particularly the role of affirmation, rather than on the form of the known.

The methodology employed in this movement can be most simply explained in terms of a two-step process, *reduction* and *deduction.*[23] Reduction begins with the immediate data of consciousness and uncovers the conditions and presuppositions implied in them. It begins with the thematically known, i.e., something that can be articulated, and moves toward what must already obtain as the condition of the possibility of this or any other particular act of knowing. Different types of conditions may be distinguished. *Ontic*

22. Immanuel Kant, *Critique of Pure Reason,* trans. N. K. Smith, 2nd ed., (London: Macmillan, 1934), p. 59.
23. For a more detailed explanation of this transcendental methodology see the introduction to E. Coreth, *Metaphysics,* trans. J. Donceel (New York: Herder & Herder, 1968), pp. 31–44.

conditions are those whose existence is presupposed by the act of knowing, but which do not enter as a constitutive element and hence are not co-affirmed in the act of knowing, e.g., the existence of light to see colors. *Logical* conditions are those whose knowledge is presupposed and implicitly affirmed in an act of knowing, e.g., in saying anything, I implicitly affirm my knowledge of the language used. *Transcendental* conditions are like logical conditions in that they are co-affirmed in the act of knowing, but like ontic conditions in that they determine the act of knowing rather than the content of the known. These are the crucial conditions. If the transcendental conditions of the act of knowing can be uncovered by reductive analysis and thematized, then one has a basis for the second step, the *deduction* of the nature and conditions of whatever can be known in accord with the transcendental conditions of the possibility of our knowing.

This general methodology has been implemented in different ways, depending in part upon the aspect of knowing that serves as the point of departure for the reduction to the *a priori* conditions of the possibility of knowing. Thus Maréchal focused on judging, Rahner on the reflective return to experience preceding judgment, and Coreth on the process of questioning. Lonergan's development is more complex but more pertinent to the present problems because of his detailed analysis of contemporary scientific thought. Lonergan's interest is, not so much in the thought as a developed product, but in the thinking, for it is here that the basic structure of human cognition is manifested in its most developed form. Lonergan's cognitional analysis proceeds from description toward structures and concludes that any complete act of human knowing is structured of three components: experience, understanding, and judgment. Experience, basically perception, supplies the material for understanding. Understanding has two steps: insight, the grasp of a potentially relevant unity or relation in the data of experience; and expression, or the articulation of this insight in the form of a concept, definition, or hypothesis. To affirm such a hypothesis as true of reality, one must have a grasp of evidence sufficient to warrant assent. This is obtained in the reflective understanding preceding judgment, a reflection that grasps the hypothesis, its evidential warrant, and the capacity of the mind to know reality. Judgment, accordingly, is also a two-step process of reflection and affirmation.

This reduction through cognitional analysis yields the structural elements present in any act of knowing, but does not of itself supply a basis for transcendental deduction. Lonergan finds such a basis in self-affirmation. The proposition, "I am a knower," must be affirmed, for to deny it would be to contradict the content of the denial by the very dynamism implicit in uttering it. In this self-affirmation one has an essential link between knowing and being and a basis for a further affirmation of being. The implementation of this begins with the horizon of being, the other pole of the I-world polarity, and builds on a heuristic anticipation of what it is to know being. Being is what is known or to be known through intelligent grasp and reasonable affirmation. Although the knowing has a three-part structure, it is essentially one act. Whatever is or can be known in this way must have a structure isomorphic to the structure of knowing. Unless there were such an isomorphism, understanding and judgment could not add the distinct contributions to knowing that they do, in fact, make.

Through a heuristic anticipation of fully explanatory knowledge, one may define 'potency' as the component of proportionate being (i.e., being proportionate to man's way of knowing) to be known by an intellectually patterned experience of the empirical residue. That is, potency is isomorphic to experience only when experience is considered as a constitutive part of full knowledge. Similarly, 'form' denotes the component of proportionate being to be known by understanding things fully in their relation to one another. Finally, 'act' denotes the component of proportionate being to be known by uttering the virtually unconditioned "Yes" of proportionate judgment. As the three levels of cognitional activity yield a single knowing, so potency, form, and act constitute a single thing, since one and the same thing is experienced, understood, and affirmed.

This general isomorphism between the structure of knowing and the structure of being admits of two parallel developments because of two different forms of explanatory knowledge. Through the abstract laws of science, we know things as they are related to other things; through knowing a thing precisely as a thing, we affirm it as an identity, unity, whole. Each requires a corresponding potency, form, and act, called 'central' for knowledge of things as units and 'conjugate' for knowledge of things as related. The latter pertain to science and admit of further subdivision in accord with Lonergan's division of scientific understanding into classical, statistical, genetic, and dialectical. Here we are more interested in the methodology and its epistemological underpinning than in the details of their development by Lonergan.

The relation between such a development of critical metaphysical realism and the problem of scientific realism is open to different interpretations. Heelan, in the article included here, argues that only such a critical realism supplies a rational basis for a justifiable affirmation of the virtually unconditioned, or of a proposition whose conditions of assertibility are fulfilled. Yet the methodology itself prescinds from the content of scientific, or any other, knowledge. The basis from which the metaphysical structure of being is deduced is the set of *a priori* transcendental conditions of knowing attained by reductive analysis. In determining the ultimate structure of reality, whatever science has to say drops out as irrelevant.[24]

If one grants the reality of the critical problem and the necessity of some sort of transcendental deduction in the establishment of a critically justifiable metaphysics—concessions that will be reconsidered at the conclusion of the Introduction—one may still ask whether the type of development sketched above is adequate to the task. Elsewhere I have attempted to probe the cogency of Lonergan's argumentation.[25] Here I wish to examine one aspect of the general methodology that bears on the problem of scientific realism, particularly on the ultimate exclusion of the contents of scientific theories. It is the problem of thematizing.

The logic of the argument sketched above indicates that the basis for such a metaphysical deduction, as in Lonergan's argument from

---

24. For many this exclusion may supply sufficient grounds for rejecting or ignoring the whole movement. It should be noted, however, that a similar objection could just as easily be brought against many philosophers in the empirical tradition. The ultimate ontic commitments come, not from the content of scientific theories, but from an anticipated logical reconstruction of scientific thought. In this case the structure of first order propositional logic plays a role similar to that played by the transcendental conditions of knowing in critical realism.

25. E. MacKinnon, S.J., "Understanding According to Bernard J. F. Lonergan, S.J.," *The Thomist*, 28 (1964): 97–132, 338–72, 475–522.

isomorphism, is the transcendental *a priori* conditions (or structures) of knowing uncovered by reductive analysis. This indication, however, is misleading, for the actual basis for deduction is the thematic expression of these conditions, or elements. The role of expression is not stressed, for the thematization is generally considered to be unproblematic. It is, unfortunately, unproblematic only for those who ignore or slight the very real problems raised by the analytic tradition. The transcendental reduction-deduction argumentation implicitly presupposes that the meaning of the key terms used in the deductive process is essentially determined by the entities, mental states, or whatever referred to. Thus if one wants to know what "insight" really means, he is encouraged to reflect on the insights he has himself experienced.

Meaning has been an abiding concern of the analytic tradition since its inception. A conclusion that has gradually won general acceptance is summarized in the proposition that the meaning of a term is not a term-thing relation.[26] Both meaning and reference are essentially system-dependent in the manner discussed earlier. This caution is particularly significant in clarifying the meaning of terms referring to mental acts or states.[27] If this contention is accepted— we refer to the sources cited for the pertinent arguments—then the transcendental deduction depends, not only on the prior reduction, but also on the conceptualization of reality in general, and mental activities in particular, implicit in the language used. Such a dependence cannot be avoided. It can be handled critically only if one has already, or at least concurrently, made an analysis of the role

26. See G. Ryle, "The Theory of Meaning" in C. A. Mace, ed., *British Philosophy in the Mid-Century: A Cambridge Symposium* (London: Allen & Unwin, 1957), pp. 239–69.

27. The difficulties such terms present have been extensively discussed since the publication of Ryle's *Concept of Mind* and especially Wittgenstein's *Philosophical Investigations* stressed the linguistic problems involved in referring to mental acts or states. The most acceptable answer seems to be that worked out by Geach in *Mental Acts,* pp. 75–78, 101–10, 124–27, and by W. Sellars in "Empiricism and the Philosophy of Mind," pp. 174–96. The core of the solution is the idea that terms are predicated of mental states or acts by analogy with overt speech. Thus a concept is understood as an inner word, and a judgment as the inner analogate of an affirmed proposition. This analogical extension of language depends on a theory of mental acts, albeit a low level theory long since assimilated into ordinary language.

of such conceptualizations and found a means of fitting this analysis into the operative methodology. If a means is found for systematically including this unavoidable conceptualization, then need others, such as the scientific world view, be excluded? We shall return to this problem in the conclusion of this introductory essay.

Systematic Analysis of Scientific Realism

The preceding survey shows that a doctrine of realism has tended to play a dual role in the interpretation of science. On one hand, the operative tradition in the practice of science almost inevitably favors some form of realism. Scientists attempt to understand and explain the real world. Their theories are developed, modified, and, when necessary, abandoned through a dialogue in which man proposes and nature disposes. On the other hand, this operative realism raises problems that render suspect any dogmatic assertion of a doctrine of realism. Too many theories have been interpreted as blueprints of the structure of physical reality only to encounter reversal and eventual rejection.

We wish here to present an interpretation of the nature and scope of the realism implicit in standard scientific thought. Since this operative realism is, as will be argued, essentially a transformed extension of common-sense realism, we will begin with a brief analysis of common-sense (or ordinary language) realism before analyzing scientific realism. Finally, we shall consider the relations between this operative realism and various positions affirming, denying, or qualifying a doctrine of philosophical realism. The positions considered will be essentially those discussed earlier, but they will now be treated systematically rather than historically.

Common-Sense Realism

Realism, in the simplest sense of the term, is a belief that the world exists as known independently of our knowing it. That such a belief underlies our ordinary activities and judgments is so clear that anyone not acting in accord with it, i.e., acting as if unseen objects do not exist or as if the experienced world were a deceptive chimera, would be regarded as psychologically abnormal. The point at issue is not the existence and efficacy of such a belief but its scope and the limits of its validity.

'Common-sense' is a broad and rather vague term referring to values, viewpoints, and judgments shared by a culture. Common-sense realism so pervades our thought and speech that it is difficult to isolate it and give an objective appraisal. If we judge the validity of common-sense realism by the criteria this realism itself supplies, our judgments will necessarily manifest an affirmative bias. If we attempt to appraise it from a different vantage point, i.e., in accord with criteria drawn from science or philosophy, our appraisal will necessarily manifest a negative bias, since we have already accepted another system of knowledge as superior to common-sense knowledge.

The situation, however, is far from hopeless. Converging methods of analysis have succeeded to a considerable extent in making explicit the structure and functioning of common-sense realism. Our interest here focuses on the view of reality implicit in shared common knowledge and ordinary language. Rather than attempt an independent development of this, we shall simply outline the salient features in Strawson's development.[28] Strawson argues from the use of language in intersubjective discourse to the conditions of the possibility of such usage. In normal conversation the speaker is able to refer to particulars such as persons, objects, and events that the hearer can identify even when the particulars referred to are not present. This can only happen if both speaker and hearer share a common space-time framework that includes both them and these particulars.

A basic condition for the possession of such a unified spatio-temporal scheme is the ability to identify some particular as the same individual encountered on previous occasions. This entails acceptance of the fact that objects perdure while unobserved. Some particulars, such as the assassin of Lincoln, are referred to only relatively. Yet, all identification cannot be relative, or reference would have an essential ambiguity that it does not, in fact, have. Accordingly, there must be a class of basic particulars. From the point of view of identification through language, these particulars are ma-

28. F. Strawson, *Individuals: An Essay in Descriptive Metaphysics* (Garden City, N.Y.: Doubleday, 1963), pp. 2–49.

terial bodies with 'persons' accepted as a special, irreducible type of material body possessing both physical and mental characteristics.

What emerges from such an analysis is called a 'descriptive metaphysics'. It is not intended as an account of the world as it really is, but as an explication of the conceptualization of reality implicit in ordinary language. It may be, as many have argued, that radically different languages, e.g., non-Indo-European languages, contain a different descriptive metaphysics. But Strawson's schema does contain, I believe, the core of our common-sense view and the conceptualization of reality from which science actually emerged.

It would be possible to argue to the same conceptualization as the core of a common-sense or ordinary language view from other perspectives. If, for example, one reinterprets an Aristotelian philosophy of nature by carefully distinguishing between empirical arguments and conceptual analysis, then much, though not all, of this philosophy is seen to be an explication of a conceptualization implied in ordinary language.[29] Phenomenological analysis can yield a roughly similar view. Thus Merleau-Ponty begins phenomenological analysis with description of the world.[30] The world is not just a collection of objects, but a milieu in which objects are related to me through my actions of knowing, desiring, etc. The referential center of this milieu, or horizon, is the lived body. Emerging from this description is the characterization of the world as perceived prior to scientific report.

Relying on these sources, we may summarize the common-sense view of reality reflected in ordinary language as follows. The world is an interrelated collection of objects that are extended in space and perdure through time. These objects are endowed with both primary (e.g., extension, shape, etc.) and secondary (e.g., color, taste, etc.) qualities that are real properties of bodies. The categorization of such properties has a natural ordering, reflected in

29. On this point see E. McMullin's article "Matter as Principle" in his book *The Concept of Matter* (Notre Dame, Ind.: Univ. of Notre Dame Press, 1963), pp. 169–208 and Pierre Aubenque, "Physique Aristotelicienne et Langue," *Archives de Philosophie*, 31 (1968): 25–34. Aristotle's analysis of change is, in large part, an explication of the way we speak about change. If a change occurs, it will be describable by statements of the form, "S is non-P" (before the change) and "S is P" (after the change). Such descriptions require three terms that can be labeled 'subject' (S), 'form' (P), and 'privation' (non-P).

30. Maurice Merleau-Ponty, *Phénoménologie de la Perception* (Paris: Librairie Gallimard, 1945).

Aristotle's table of categories: substance, quantity, quality, etc. Thus, "Whatever is colored is extended" is an *a priori* proposition of conceptual necessity, while its converse is not, because secondary qualities (or proper sensibles) are conceived of as adhering in substance through the mediation of primary qualities (or common sensibles). Things, or substances, are of different ontological types, with persons as a unique class of which both mental and corporeal attributes are predicated. Among persons I have a unique role. My bodily location anchors the space-time framework and establishes the horizons making the world 'my world'. It is also a shared world, a fact that is best explained by accepting language as a symbolic vehicle that I assimilate by adapting to the public meaning of words and then use to structure my observation reports. So dominant is this structuring effect that any attempt to bracket it and develop a pure description of the world as immediately perceived encounters severe, perhaps insuperable, difficulties.

Common-sense realism, in its varied forms, can be used as a basis for the interpretation of scientific theories in various ways. We will indicate appropriate strategy arguments under the headings: epistemological, methodological, and ontological.

Epistemological Interpretation

All knowledge of the external world begins with, and is ultimately based on, sensation. Even if one acepts sensation and intellection as irreducibly different types of knowledge, it is the former rather than the latter that is constitutive of physical objects. The basic reason for this is that sensation as distinguished from perception is essentially the *passive* reception of stimula due to external causal agents, while intellection is the *active* construction of objects by the mind. Scientific theories, accordingly, must be given a physical interpretation by reducing them to their basis in sensation. This line of reasoning may be found in its purest and simplest form in the phenomenalism of Ernst Mach and Karl Pearson and in some developments of sense-data theories, such as that of Bertrand Russell at one stage of his varied intellectual career.

A fundamental difficulty encountered by any phenomenalist interpretation of science is that scientific verification depends on observation reports, which are public and objective, rather than on simple sensation, which is private and subjective. The logical positivists tried to counter this difficulty by theories of scientific language in which some form of sensation reports supplied an interpretative basis. Increasing epistemological sophistication led to a transition from protocol sentences to observation reports to the idea of developing an empiricist language. Yet each attempt encountered the difficulties already considered in discussing bottom-of-the-page interpretations of scientific theories. They minimize the role of theories, distort the interpretation scientists give to theoretical entities, and ultimately rely on the epistemological myth of the given.

Methodological Interpretation

One might forego any attempt to reduce the verifiable elements in scientific theories to a basis in sensation and yet hold that common-sense realism is the ultimate interpretant of science. Although neither Strawson nor Merleau-Ponty has developed an interpretation of physics, each insists in his own way that the common-sense view of reality is more fundamental than any scientific view that seeks to replace it.[31] Such an approach to the interpretation of science has recently developed in a more systematic way by Körner.[32] Körner insists on a complete, logical dichotomy between empirical discourse (essentially the type of system Strawson has clarified) and theoretical discourse. The ontic commitments of scientific theories are to ideal objects, and these theories are interpreted by a partial identification of the objects, predicates, and propositions of theoretical discourse with their empirical counterparts.

In a different way, both Bohr and Heisenberg have insisted on the primacy of a common-sense or ordinary language view of reality in any attempt to give a physical interpretation to scientific theories. As Heisenberg explained it:

31. See Strawson, "Introduction," pp. xii–xvi, and Merleau-Ponty, "Avant-Propos," pp. ii, iii. A more detailed defense of Strawson's position may be found in his article, "Carnap's Views on Constructed Systems versus Natural Languages in Analytic Philosophy," in P. A. Schilpp, ed., *The Philosophy of Rudolf Carnap* (La Salle, Ill.: Open Court, 1963), pp. 503–18.

32. Körner, *Experience and Theory: An Essay in the Philosophy of Science* (New York: Humanities Press, 1966).

Furthermore, one of the most important features of the development and analysis of modern physics is the experience that the concepts of natural language, vaguely defined as they are, seem to be more stable in the expansion of knowldge than the precise terms of scientific language, derived as an idealization from only limited groups of phenomena. This is in fact not surprising since the concepts of natural language are formed by the immediate connection with reality; they represent reality.[33]

Quoting Heisenberg along with ordinary language analysts and phenomenologists might give the misleading impression that he also wishes to restrict the explanatory scope of scientific theories. What he actually does is indicate that the word 'reality' is an ordinary language term that can be extended to scientific theories only in an indirect way governed by the principles of complementarity. Yet, the fact remains that the common-sense realism of ordinary language plays a decisive role in the mode of interpreting scientific systems by Bohr and Heisenberg.

Ontological Interpretation

Here we shall focus on a line of argumentation developed by scholastic philosophers and, for a different purpose, by G. E. Moore. There are many statements made within ordinary language that are true and known with certainty to be true. "There exists at present a living human body that I know to be my body," and "This observed material object is here" are examples of statements that I, or other persons, can affirm with such certitude that any system of thought that calls their truth into question is automatically suspect.[34]

If truth is defined as the conformity between what is said and what is, then statements already known to be true must conform to reality. Unless this paper existed and really were white, the

33. Werner Heisenberg, *Physics and Philosophy: The Revolution in Modern Science* (New York: Harper & Row, 1958), p. 200.
34. See G. E. Moore, "A Defense of Common Sense" in R. Ammerman, ed., *Classics of Analytic Philosophy* (New York: McGraw-Hill, 1965), pp. 47–67. For a similar development within the scholastic tradition see Reginald F. O'Neill, S.J., *Theories of Knowledge* (Englewood Cliffs, N.J.: Prentice-Hall, 1960), chaps. 4–10.

statement "This paper on which I am now writing is white" would be false. Immediate sensation coupled to a basic understanding of the meaning and referents of the terms 'paper' and 'white' conclusively proves that the statement in question is true. One must conclude, accordingly, that being white is a natural property of bodies. More generally, what is called a 'descriptive metaphysics' is not simply an explication of a conceptualization implicit in a language system, it is essentially a description of reality as it is. Bodies *are* mobile, enduring objects endowed with primary and secondary qualities that are real ontological determinants of these bodies. Any scientific system that seems to call this into question, such as the wave theory of light, must be considered a hypothetical system of knowledge that speaks of theoretical constructs rather than of reality as immediately experienced.

Scientific Realism

Many readers will not find the preceding arguments conclusive. But before returning to the linguistic and epistemological problems raised, we will do well to consider functional scientific realism, for it constitutes a conceptualization of reality that competes with the descriptive metaphysics of ordinary language. Without a competing conceptualization, there is no adequate way to distinguish between internal and external questions relative to this descriptive metaphysics.

Scientific realism is seen in its starkest form in classical Newtonian mechanism. This was basically the belief that the world exists as scientific theory pictures it. If the basic laws of physics are interpreted as laws built into the fabric of reality, something discovered rather than invented, then the conceptualization of reality proper to these systems is interpreted as the most fundamental way of understanding reality itself. The preceding historical survey indicated the reasons leading to the abandonment of this type of simple realism. The point is most clearly seen in Duhem's distinction between the representative and explanatory parts of physics. If only the former, the formal structure, is accepted as a constitutive part of physics, then the conceptualization of reality that is the explanatory part of physics may be disregarded on the

grounds that, despite its psychological role in discovery and interpretation, it does not play a necessary role in explanation.

This reaction against a naive realism coupled to an empiricist
epistemology led to a radical devaluation of the role played by a
conceptualization of reality in a physical theory. This was indicated
earlier when discussing bottom- and middle-of-the-page interpretations of scientific theories. It can be seen more clearly by analyzing what is involved in a formal approach to the interpretation of
scientific theories. In such a view a scientific system is explained
in terms of three components: an abstract calculus, an observational framework, and correspondence rules linking the two. If the
abstract calculus is considered a formal uninterpreted system, then
it receives an interpretation through the correspondence rules that
link terms in the theoretical language (or abstract calculus) with
physically significant terms in the observational language. Thus, one
would link T with temperature as measured, etc. If the theoretical
language receives an interpretation only through such correspondence rules, then it cannot be said to contain a conceptualization
of reality independent of that proper to the observational language.
The observational language expresses physical reality, while the
theoretical language is essentially an inference mechanism leading
from one set of observation statements to another.

Recent reactions against such purely formal interpretations of
scientific systems have led to a top-of-the-page interpretation and
to something of a return to realism, which is evident in Sellars's
paper. Here a scientific theory must be considered a conceptualization of reality, at least in the minimal sense of specifying the
entities discussed within the theory. This mode of interpretation is,
I believe, correct. But its implementation is often vitiated by the
artificiality of the systems considered. The criticisms that follow
apply more appropriately to the work of Quine [35] and especially
Scheffler [36] than they do to Sellars. These theories of science focus,

35. See particularly W. V. O. Quine, *Word and Object* (Cambridge, Mass.:
M.I.T. Press, 1960), esp. chaps. 5 and 6, where scientific discourse is regi-
mented into the austere framework of first order sentential calculus.

36. Israel Scheffler, *The Anatomy of Inquiry: Philosophical Studies in the
Theory of Science* (New York: Knopf, 1963). The critical difficulties in
Scheffler's interpretation are discussed in detail in my review article "Epistemological Problems in the Philosophy of Science," *The Review of Metaphysics,*
22 (1968): 112–37, esp. 114–24.

not on actual scientific theories, but on the logical form that such theories would have if they were rationally reconstructed in accord with logicians' criteria. In practice, this means assuming a first order sentential calculus or a system of pure logic extensionally interpreted and supplemented by special axioms for different theories. The question of the entities postulated by the theory is thus transposed into a question of the entities, or the ontic commitments of the theory, whose names serve as values for bound variables in the axioms. This often degenerates into another round of the nominalist-platonist controversy when the question of whether classes (or properties) should be allowed as values of variables. Regardless of what one thinks of this venerable controversy, it is clearly a controversy about the foundations of logic rather than physics and has little to do with the question of whether atoms, particles, and the other theoretical entities postulated by physical theories should be considered existentially real or merely theoretical constructs.

To implement this new approach to scientific realism, we should develop a method of making explicit the ontic commitments actually operative in physical theories, rather than those that might be proper to a structural description presumed to be applicable to all theories when rationally reconstructed. Here recent studies in the history of science by Kuhn, Jaki, Hanson, and others offer some guidelines with their stress on the role of conceptualization in physics, a role that is particularly evident in an examination of conceptual revolutions. These studies do not, however, clarify the relationship between these history-centered analyses of theories as conceptualizations of reality and logic-centered analyses of theories as formal axiomatic systems. In what follows we hope to clarify this relationship and in so doing make explicit the nature and limitations of functional scientific realism.

As indicated earlier, language cannot serve as a means of describing, relating, narrating, or explaining unless it contains some conceptualization of the reality referred to. This is true, not only for ordinary language, but also for the specialized and transformed extensions of ordinary language that function in scientific theories. Linguistically, contemporary science grew out of Indo-European languages. Attempts to understand puzzling phenomena led to insights expressed through new concepts. Formal concepts are interiorized words whose meanings are determined by the way they function in language. An insight or a sudden new way of understanding a problem can give public meaning to a concept only by controlling the way the correlative term is used, for instance through definitions, explanations, or examples. Accordingly, a study of the way language is used is a key to understanding the way reality is conceptualized.

A couple of historical examples may help to clarify this. Both Descartes and Newton considered inertia a basic concept in mechanics, and both, building on Galileo's work, regarded inertia as the tendency of a body at rest to remain at rest or, if in motion, to continue in motion in a straight line. Yet the two conceived 'body' differently. For Descartes, extension was a basic property of material bodies. 'Inertia', accordingly, was explained as the product of the volume of a body and its speed. However, the mathematical theory of collisions based on these definitions proved unsuccessful. Newton conceived of mass rather than extension as the basic property of material bodies, defined 'inertia' as the product of mass and velocity, and thus developed a highly successful mechanics.[37]

From Newton's mechanics emerged the new conceptualization of material reality described in the preceding historical summary. It had a basic continuity with the descriptive metaphysics of ordinary language and with the earlier Aristotelian physics in that material reality was conceived of as a collection of bodies with properties, rather than as events or as a monistic substance. The properties that were basic, however, were mass, inertia, and gravitational attraction, while the secondary qualities of ordinary language were disregarded.

Before considering further examples, we should consider the relation between such a conceptualization of reality and a mathematical system. Extension and its relation to ordinary arithmetic affords a simple example. 'Length' is spoken of as a property of a body. By introducing higher order predicates, we may speak of the properties of length in terms of its behavior under different operations. One may simply juxtapose lengths in a straight line or join them at an

37. This divergence between the Cartesian and Newtonian analyses of inertia is examined in more detail in Richard Blackwell, *Discovery in the Physical Sciences* (Notre Dame, Ind.: Univ. of Notre Dame Press, 1969), chap. 1.

angle giving an area. These combinations are associative $[(a \cdot b) \cdot c = a \cdot (b \cdot c)]$ (where '·' replaces either '+' or '×'), commutative $(a \cdot b = b \cdot a)$, and distributive $[a \times (b + c) = (a \times b) + (a \times c)]$. These second order predicates of material bodies, defined with respect to the operations of linear and rectilinear joining, are formally identical with the first order properties of numbers. This isomorphism gives a basis for the applicability of arithmetic to mensuration.[38]

Basing the correspondence between a physical language and a mathematical formalism on the isomorphism between second order predicates applied to physical entities (or predicates of predicates) and first order properties of mathematical entities allows for the possibility of preserving a mathematical formalism while changing the underlying conceptualization. This is illustrated by the development of thermodynamics.

A science of thermodynamics first became possible when Fahrenheit and Celsius established temperature scales early in the eighteenth century. The initial attempts to develop appropriate mathematical laws failed chiefly because no operative distinction between degrees of temperature and quantity of heat existed. This distinction, as well as the notions of specific heat and latent heat, was supplied by Joseph Black. At this period heat was regarded as a fluid with chemical as well as thermal properties. The conceptualization became clearer after Lavoisier rejected the phlogiston theory of combustion and in his *Traité élementaire de chimie* (1789) listed caloric as a basic element. Caloric was believed to be an indestructible, uncreatable, weightless, and elastic fluid existing in either sensible or latent form. Its particles repel one another, thus explain-

---

38. This idea of the role of second order predicates was derived from W. Sellars, "Scientific Realism or Irenic Instrumentalism." The same idea had been developed on the basis of Lonergan's theory of knowledge in an unpublished article by T. Linden, S.J. It may be argued that such a correlation is highly artificial and not operative in normal scientific procedure. Piaget and his coworkers, however, have clearly established the role that higher order predicates play in any development of knowledge. Thus, a growing child masters the concept of volume of a fluid only when he realizes that the quantity of water in a container is invariant under changes in shape, e.g., pouring water from a tall, thin glass into a short, wide glass. A first order property of bodies, here quantity, is not properly assimilated until *its* properties, e.g., invariance under certain transformations, are also assimilated. See B. Inhelder and J. Piaget, *The Growth of Logical Thinking from Childhood to Adolescence* (New York: Basic Books, 1958).

ing the diffusion of heat. They are attracted by particles of ordinary matter, the magnitude of the attraction being different for different substances and for different states of the same substance.

This conceptualization of heat was similar to the conceptualization of matter that had proved successful in Newtonian mechanics. Matter was pictured as a collection of mobile, enduring particles with real primary qualities, specifically mass and gravitational attraction, and no secondary qualities. Caloric was similar except for different primary qualities. This model served as a successful basis for the mathematical treatment of heat exchange.

If one accepts caloric as an explanation of what heat is, then within this framework, or physical language system, one can examine further implications of this conceptualization. Count Rumford did so in trying to explain frictional heat in terms of caloric theory. He concluded that the extreme quantity of heat produced in boring a cannon was incompatible with the conservation of caloric. The problem then was to develop a new conceptualization of heat that would preserve the valid mathematical laws established on the basis of the caloric theory. For our purposes, the subsequent work of Mayer, Joule, Maxwell, and others can be telescoped into three key steps. First, the thesis that heat is a form of energy rather than a fluid was established by Joule's demonstration of the mechanical equivalent of heat. The law of the conversion and interconvertibility of energy developed by Mayer, Clausius, and Kelvin preserved and generalized the laws hitherto explained by the indestructibility of caloric. Finally, Clausius's revival and Maxwell's extension of the theory that a gas is a collection of molecules in motion and that temperature is the phenomenological equivalent of the average kinetic energy of these molecules supplied a mechanical basis for thermodynamics, now considered a phenomenological science.

The conclusion suggested by these considerations is that the two-component view of scientific theories is correct in principle, but that both components have been misinterpreted. The dichotomy between an observational language and a theoretical language was determined more by prior epistemological commitments than by an analysis of the way scientific theories develop and function. Accepting as fundamental a division of all meaningful statements into analytic and synthetic and a program of reductionism, philosophers felt compelled to ground factual statements in observation and formal statements in the principles of logic. The truncations and distortions due to this procrustean effort are becoming ever more evident.

It is more meaningful to speak of these components as a 'physical language' and a 'mathematical language'. The physical language embodies a conceptualization of a domain of reality and serves as an indispensable vehicle for referring to objects in that domain, or to aspects of it, and for saying something about these objects or aspects. The conceptualization of reality implicit in this language can be explained as an extension of the descriptive metaphysics of ordinary language modified by successive transformations that may be briefly summarized.

The core of the Aristotelian-scholastic philosophy of nature *is* an explication of the descriptive metaphysics of ordinary Indo-European languages. Reality is conceived of as a collection of mobile, extended, relatively enduring bodies that are endowed with distinctive properties and characteristic activities and are set in a unified space-time framework. Bodies are conceived of as continua with properties having a certain ontological ordering: substance, quantity, quality, relation, action, etc. The order of intelligibility is from substance, down. What this means is an analysis of the conceptualization implicit in the way we speak of bodies. Thus we speak of extension only in terms of something extended, of shape only in terms of an extended something having shape, etc. This order of intelligibility supplied the basis for the Thomistic gradation of beings in terms of the perfections of their natures. The perfection is manifested by the degree in which their characteristic activities transcended a dependence on matter.

Newtonian mechanism preserved this general structure, but modified some categorical features. Bodies were regarded as substances with primary qualities while secondary qualities were essentially relocated in the attempt to reduce them to primary qualities. The basic primary qualities were extension, mass, inertia, and gravitational attraction. With qualities, in the Aristotelian sense, effectively removed from the hierarchical ordering, actions had to be explained in terms of quantified matter, that is, by reduction to local motion, attraction, and repulsion. This effectively undercut any basis for conceptualization of substances with distinctive natures and logically favored a program of materialistic reductionism.

This conceptualization served as the ground for a mathematical system in the manner discussed earlier, a subject to which we shall return. It also served as a paradigm for other physical theories treating light, electricity, magnetism, and heat. Twentieth-century developments in physics introduced further modifications in this conceptualization. Because quantum mechanics rejects models as pictures of reality, the role that a conceptualization of reality plays in this theory had been rather neglected until recent debates about the quantum theory of measurement brought it to the forefront. The role this plays and its relation to our interpretation of science as a two-component system may be seen by analyzing Dirac's authoritative and highly influential treatise on quantum mechanics.

> The new scheme became a precise physical theory when all the axioms and rules governing the mathematical quantities are specified and when in addition certain laws are laid down connecting physical facts with the mathematical formalism, so that from any given physical condition equations between the mathematical quantities may be inferred and vice versa.[39]

Here Dirac interprets a theory as containing a physical component, a mathematical component, and rules connecting the two. In the physical system the basic distinction is between systems and states. Thus a proton plus an electron may form one system with various states. The basic properties or dynamic variables of the system may be specified. The classical properties, i.e., mass, location, momentum, etc., carry over but with new restrictions specified by the superposition principle (which implies the indeterminancy principle). This is a restriction on the acceptable way of speaking about a system. Thus a precise specification of momentum precludes a precise specification of location.

The way in which the superposition principle serves as a bridge to mathematics requires clarification. The principle asserts that the state of a quantum mechanical system may be regarded as a superposition of two or more states in a way that cannot be conceived of in terms of classical ideas. Thus a photon with a polarization in a definite direction may be considered a superposition of

---

39. A. M. Dirac, *The Principles of Quantum Mechanics*, 4th ed. (Oxford: Clarendon Press, 1958). This method of interpreting science will be given a more technical analysis in my article "Ontic Commitments of Quantum Mechanics," *Boston Studies in the Philosophy of Science* (to be published).

two photons with polarizations 45° to the right and left of the given direction. A classical analogue would be to say that a man weighing 175 pounds can be considered a superposition of two men weighing 150 and 200 pounds. An attempt to weigh the man might catch either of the two superimposed men, but if we weigh him often enough the average will be 175.

The classical analogue is not only absurd, because weighing does not produce such results, but also unintelligible, because weighing cannot produce such results. The classical conceptualization of bodies does not admit of such indeterminateness in primary qualities. Yet the quantum mechanical conceptualization must admit this because crucial experiments cannot be explained on any other basis. Earlier the relation between the physical language and the mathematical language was explained by saying that a mathematical quantity represents a property of a system. But the question of which mathematical entities are appropriate depends on a higher order correspondence between the properties of the physical property (or second order predicates applicable to the physical body) and the first order properties of the mathematical entity. This is essentially the criterion Dirac uses.

The basic correspondence must be between states of a system and a class of mathematical entities that represent these states. But the properties of the states are specified by the superposition principle. The appropriate mathematical entities must be such that a given unit is equivalent to a superposition of two other units, a characteristic of vectors. Granted rather complicated vectors, axioms and rules governing their behavior must be specified as the basis of the mathematical representation. These need not be considered here.[40]

Before returning to the role of conceptualization or descriptive metaphysics, we should clarify the significance attached to correspondence rules. They do not relate an uninterpreted formal theory to a theory-free observation language. Rather, their basic function is to establish a correlation between properties (loosely conceived) specifying a physical system and appropriate mathematical entities.

40. There is some doubt about the adequacy of a Hilbert space representation for relativistic quantum mechanics, although a better representation has not yet been developed. However, there is no doubt about its adequacy for non-relativistic quantum mechanics, the subject under consideration.

This correlation is controlled by a formal identity between second order physical predicates and first order mathematical predicates. Acceptance of this higher level correspondence as controlling has further implications. This may be preserved while the first order physical predicates, or those that characterize reality rather than the way we speak about reality, are changed.

Thus the caloric theory of heat was abandoned in favor of an energetic theory of heat, but the relation of mathematical laws to physical systems could be retained because the essential higher order properties remained. The conservation of caloric was reinterpreted as a special case of energy conservation, and this conservation law required an additive arithmetic. More generally, the history of many physical theories may be seen as a peculiar fusion of progress and revolution. The progress is in more accurate and more extensive mathematical laws. The revolution is in the conceptual underpinning of these laws. Gravity is a clear case. Galileo's laws for falling bodies presupposed that gravity was a constant force field surrounding the earth. Newton's laws are uniquely compatible with a view that allows of action at a distance, although Newton himself refused to accept this on philosophical grounds. Einstein conceived of gravity as the warping of space in the vicinity of massive objects. These three—a constant force field, action at a distance, and warped space—are not mutually compatible conceptualizations; none are compatible with the suggested quantum explanations of gravity. Yet the mathematical laws are successive approximations, in the sense that each reduces to its predecessor in the domain and within the limits of validity of the predecessor.

The conceptualization of reality, or of a domain of reality, implicit in the thought and language of a physical theory, plays a functional rather than a philosophical role. Yet, shaky as it may seem, it is the foundation for the functional realism operative in the tradition and practice of science. This does not mean that the type of conceptualization considered is completely or uncritically accepted by practicing scientists. It is not. But it does play an indispensable role in the formulation of statements accepted as true; this is the crucial point.

The ultimate justification of common-sense or ordinary language realism rests on the realization that within ordinary language one can make statements that are both true and known with certainty to be true, e.g., "This paper is white." A similar justification can be given for scientific realism. Even discounting tautologies, statements true by convention, and other disputed cases, the scientist can make factual statements that he reasonably accepts as true. The *Handbook of Chemistry and Physics* is a reservoir of such statements: e.g., "The electrical resistance of iron is greater than silver"; "The normal helium atom has two electrons," etc. The basic argumentation is the same in both the ordinary language and the scientific cases. To accept as true a statement such as "The electron has a unit negative charge" entails a commitment to the real existence of electrons as things with such properties as having electrical charge.

Our consideration of common-sense realism concluded with a presentation, but not an evaluation, of such a proposed justification of realism. An evaluation obviously depends on a critically elaborated theory of truth. Here I shall simply apply some of the principles that I have attempted to justify elsewhere.[41] The correspondence theory of truth correctly expresses, I believe, the intentionality of a truth claim. That is, in affirming a proposition as true, one is implicitly claiming a correspondence between an expressive unit in a symbolic vehicle, language, and something extra-linguistic, such as a state of affairs. This particular correspondence depends on the general correspondence between language and reality, something that cannot be completely expressed in language. Accordingly, the attempt to clarify the significance of a truth claim focuses on the middle distance, on the presuppositions and entailments of the proposition, and on the reasons that warrant asserting it as true.

Let us return to a typical observation statement, "This paper is white," and consider what is entailed in accepting it as true. Before learning to use color terms, one must have the ability to make sensory discriminations, e.g., to see the difference between white objects and black or brown objects. Granted this, one learns to use the term properly by hearing it applied to standard members (white objects) and denied of standard non-members. If we now compare the sensory experience with its linguistic correlate, we discern an isomorphism embedded in qualitative differences. The

41. Edward MacKinnon, *Truth and Expression: The Hecker Lectures, 1968* (Paramus, N.J.: Paulist/Newman Press, 1971).

qualitative aspects of seeing white, hearing C# on a violin, tasting a rare steak, smelling a fresh rose, etc., are not reproduced in language. These qualities are sensual rather than conceptual. On the basis of such experiences, however, we can distinguish objects by their qualitative differences, and this discrimination can be reproduced in language, e.g., through learning to associate different color terms with different sensations.

Just as sensed experiences have irreducible qualitative aspects, so too do the corresponding terms. The qualitative aspects here, however, are conceptual rather than sensual. The logic of color words dictates that the application of the predicate 'red' precludes the simultaneous application of 'blue', but allows 'colored'. It also entails that if something is red, then it is extended. Both 'red' and 'extended' are predicated of 'something', but the order of conceptualization is: 'something', 'extended', 'red'.

"This paper is white" could be given two different types of interpretations. First, it could be interpreted merely as an observation report. As such, it implies that any other truthful, capable observer seeing the same object under normal circumstances would make the same report. In making such a report, one uses, but need not affirm, the descriptive metaphysics implicit in the language used. Secondly, the statement could be interpreted as a substantive claim to the effect that the quality of whiteness inheres in or modifies the substance denoted by the subject term. Is such a claim justified?

The correlation between perception and language does not of itself justify attributing to reality the descriptive metaphysics implicit in the language used. This is, in Carnap's sense, an external question calling for a judgment about, not within, the system. If this conceptualization is adequate to one's felt needs and no better conceptualization is known, then one is pragmatically justified in accepting the descriptive metaphysics of ordinary language as basic.

As it happens, more adequate conceptualizations are available. According to the wave theory of light, the molecules in this paper have the property of so interacting with incident radiation through selective absorption and reflection that the light transmitted has a fairly uniform distribution in wavelength ranging from 3970 Å to 7230 Å. Such radiation impinging on my retina produces sensations that I have learned to correlate with the term 'white'. Affirming such a conceptualization as an explanation of seeing something white does not entail denying the truth value of the observation report, "This is white," nor does it imply reducing the meaning of 'white' to 'looking white'. It simply accepts as explanatory a conceptualization of reality different from the conceptualization implicit in ordinary language.

Just as there is a partial isomorphism between perceptual discriminations and appropriate ordinary language terms, so there must be a partial isomorphism between qualitative terms in ordinary language and appropriate terms in this specialized extension of ordinary language. Thus red corresponds to 7230–6470 Å, orange to 6470–5850 Å, etc. But the wavelengths in question are conceived of as qualitative aspects of postulated theoretical entities. As Sellars points out in the article included in this book, if there is good reason for accepting this explanatory system as true, then there is equally good reason for accepting the entities it postulates, e.g., electromagnetic radiation, as real. If, however, one can in turn reduce this to a more fundamental system, e.g., quantum electrodynamics, then one is not logically obliged to affirm the ontic commitments of the reduced system.

From within a given linguistic framework one cannot precisely specify the degree to which, in affirming a proposition as true, one is also affirming the framework features that condition the meaning of the proposition and specify the way the proposition, or key terms in the proposition, refer to reality. A true proposition is an intentional absolute embedded in a conceptual relative. Scientists, conscious of competing theories and the changes that theories have undergone, are usually more aware of the relativity of conceptual frameworks than those who, whether by choice or default, are confined to ordinary language.

The implications of these considerations for the problem of scientific realism are probably clear. The scientist in his ordinary practice accepts and affirms propositions as true in a correspondence sense and as statements of what obtains in reality. He cannot do this without implicitly affirming at least the basic features of a conceptualization of reality. This is necessary to specify the objects he is speaking about, the properties, relations, etc., he is attributing to them, and his mode of reference. In this sense, science as practiced necessarily entails a functional realism. This realism is functional inasmuch as it has a definite function within science. Without it there could be no truth claims, and without truth claims there would be no science. But the fact that such conceptualizations must be used does not, of itself, warrant asserting them as true in a correspondence sense. The realism operative within the scientific tradition generates, but does not solve, a critical problem. Do such conceptualizations, or at least a consistent core developed from partially competing and partially complementary conceptualizations, represent the really real, the world as it actually is?

Plato insisted on the necessity of subordinating science to philosophy if scientific statements are to be true as well as formally correct, a contention repeated in varying forms by subsequent metaphysicians. The analyst often uses philosophy to dissolve rather than solve problems. Each, in his own way, insists that the philosopher is uniquely qualified to treat the critical problems that scientific realism generates but does not solve. In the final section we well attempt to evaluate how well philosophers have fulfilled this task.

Philosophical Realism

Both common-sense realism and the functional realism operative within the scientific tradition are pre-critical. They are based on the ontic commitments and the conceptualization of reality operative within a linguistic framework and on the pragmatic decision, whether implicit or explicit, to accept such a framework as the basis for further development. The questions considered are internal, in Carnap's sense, rather than external. An awareness of the role of conceptual revolutions and competing theories modifies this use-oriented acceptance of frameworks by suggesting the more basic epistemological question of whether any conceptual framework can represent the really real, reality as it exists independent of our knowing it.

This perenially perplexing problem is not the subject of the present study, but it cannot be altogether ignored. We will consider the relation between the functional realism already treated and different proposed methods of solving or dissolving the problem of philosophical realism. This emphasis on methodology in turn suggests the feasibility of focusing on strategy arguments supporting different developments, rather than a precise consideration of the particular positions advocated by individual philosophers. What follows, accordingly, is a simplified spectrum of some possible positions on metaphysical realism and an evaluation focused on the strategy arguments supporting these positions and on their relevance to the problem of scientific realism.

Metaphysical Realism Not a Problem

The thesis that metaphysical realism is not a problem is proposed by two radically differing groups: immediate realists, who insist that such realism is so obviously true that it cannot be seriously questioned; and anti-metaphysicians, who say that it is not a problem because it is ultimately meaningless. To the degree that immediate realism represents a refined version of common-sense realism, it has already been considered and its inadequacies noted. However the position is also supported by supplementary arguments, which, if valid, undercut the significance we have attached to the critical problem. Hence, they must be at least briefly considered.

The first contention is that the critical question can only be posed in idealistic terms, i.e., beginning with ideas and asking if the world corresponds to our ideas of it. Any attempt to answer this question and to supply a critical justification for metaphysical principles is really a misguided attempt to justify realism by idealistic principles. The realist, accordingly, should refrain from questioning the validity of his position or from attempting to justify it except indirectly by showing that any other position reduces to an untenable idealism.

The second supporting argument is the contention that knowledge is only intelligible as an interaction between mind and reality, an interaction grounded in the mind's receptivity to externally produced stimuli. The only alternatives to such an interactionist view are solipsism and pre-established harmony, and it is difficult to take either seriously. Knowledge emerges from a man-world interaction, and its validity must be judged by the degree to which knowledge reflects reality. This is a fundamental idea in Marxist philosophy.

Although each of these arguments has a valid nucleus, neither can bear the weight imposed on it. The acceptance of a man-world interactionism as the basis of knowledge is insufficient to justify immediate realism unless it can be shown that the basic categories and principles of this position represent transcriptions into language of the given of reality or immediate experience. Even if one ignores the formidable arguments that have been marshalled against this notion of givenness,[42] further difficulties remain. Any argument from givenness necessarily involves a discrimination between given and imposed elements in knowledge. This discrimination, if not to be arbitrary, requires a rationale. Admitting this entails abandoning both the claim that philosophical realism requires no prior critical justification and the corollary that metaphysics must ground epistemology. If one either refuses to admit the need for such discrimination—a position that cannot be consistently maintained—or uses the metaphysical system in question as the ground for the requisite discrimination, then immediate realism has degenerated into dogmatic realism. While this may be acceptable to some on the grounds of party loyalty, such grounds are essentially foreign to the philosophical enterprise.

The opposite extreme finds its clearest expression in Carnap's contention that realism is a pseudo-problem. This anti-metaphysics, stemming from the early positivistic theories of meaning and verification, has subsided in recent years. More significant in a contemporary context is the Wittgensteinian idea that philosophy is basically, or even exclusively, an activity of clarification. The function of the philosopher is to clarify the structure, meaning, and use of discourse in both ordinary and constructed languages. Philosophy, so conceived, is a higher order disicipline. Its proper domain is discourse about objects rather than the objects themselves. Instead of attempting to answer the question "What is the nature of X?" the philosopher should concentrate on the question "How is the word 'X' used?"

This difference is often characterized by a simple distinction between first and second order questions. But this facile distinction does not get to the heart of the problem. Suppose a metaphysical realist constructs a system, e.g., neo-Thomism, that is interpreted as a set of first order statements about the ultimate nature of physical reality, e.g., every being is composed of two really distinct constitutive principles, essence and existence. The analyst can then interpret

42. Sellars, "Empiricism and the Philosophy of Mind."

this as a proposal for an ideal language and argue that the reasons offered by the realist do not support a stronger conclusion.[43]

This is an oblique conflict rather than a direct confrontation. The analyst's questions and criticisms are metaphilosophical rather than philosophical in the metaphysician's sense of stating truths about the world. To answer that this is what the statements mean because this is what the speaker intends misses the issue, which is the contention that the arguments offered by the metaphysician can be reinterpreted without residue as a proposal for an ideal language. The only effective answer is to show either that the metaphysician's statements should be accepted as statements about the world, in accord with a theory of acceptance, or that some metaphysical statements must necessarily be interpreted as statements about the world. These two approaches will be explored in the next two sections.

Analytic Pragmatism

By suitable refinement and redevelopment it is possible to develop a metaphysics—in fact a multiplicity of metaphysical systems—from within an ordinary language framework. If all such systems are judged inadequate as a basis for a robust and critically developed doctrine of realism, then the next logical step is to examine whether the scientific view of the world can supply a more acceptable basis.[44] The type of scientific realism outlined earlier is essentially functional and not geared to answer the philosopher's questions. But if ordinary realism can be redeveloped in a consistent and systematic way that makes explicit the descriptive metaphysics it contains implicitly, then perhaps something similar can be done for scientific realism.

Previously, this was done easily and uncritically by accepting one scientific theory, Newtonian mechanics, as the system that was both fundamental, in that it reflected nature's most basic laws, and fixed,

43. See R. Roarty, "Do Analysts and Metaphysicians Disagree?" in *Proceedings of the American Catholic Philosophical Association*, 41 (1967): 39–53 and the introduction to his book *The Linguistic Turn: Recent Essays in Philosophcal Method* (Chicago: Univ. of Chicago Press, 1967), pp. 1–39.

44. For a more extended discussion of the role of the common-sense versus the scientific views of the world see W. Sellars, "Philosophy and the Scientific Image of Man" in *Science, Perception and Reality*, pp. 1–40.

in that it was judged immune to conceptual revolution. Such simple scientific realism is no longer viable. Rather, one must consciously and critically construct from the scientific world view a consistent conceptualization of reality. This is sometimes done (e.g., by Quine) by regimenting scientific theories to fit an interpretative framework structured by certain logical and epistemological criteria. However, for reasons indicated earlier, such projected, though rarely realized, reformulations of scientific theories along the lines of classical predicate logic do not do justice to the conceptualization of reality implicit in scientific theories. A more promising approach is to explicitate such conceptualizations and then seek to systematize them into a unified, coherent conceptualization of reality.

By utilizing the knowledge and theories of science, it is possible to construct an ontology within a naturalistic or pre-critical framework. The general strategy would run along something like the following lines. Knowledge, or at least its public expression, may be viewed as an interrelated set of conceptual systems ranging from the conceptualization of reality implicit in ordinary language, through the social, psychological, and biological sciences, on to the ultimate physical sciences. Some of these sciences do not involve distinctive entitive commitments: for example, thermodynamics interpreted as a study of the phenomenon of heat, or special relativity, which sets the boundary conditions that any acceptable physical theory must meet. Those that do involve such entitive commitments can be considered from a minimal point of view. Thus different theories of quantum mechanics (but not quantum field theory) implicitly involve a commitment to fundamental particles as basic units. Such commitments are essentially functional and, in principle, revisable. But revisions of substantive theories tend to carry over, though in a revised form, a central core of substantive commitments. Atomic physics, for example, has grown through a process of theory replacements. Yet a statement such as "The normal helium atom has two electrons" preserves its basic meaning and truth as one passes from the Bohr to the Schrödinger to the Dirac theory of the atom. The issue here is the descriptive metaphysics implicit in the physical language (in the sense explained earlier) of scientific theories considered as conceptualizations rather than as formal axiomatic systems.

Functional and revisable as they may be, such substantive commitments are necessary in the sense that no scientific system can be used to describe, explain, or refer to any domain of reality without at least an implicit minimal conceptualization of the entities proper to that domain together with their basic properties, activities, and relations. By broadening and denominalizing Quine's terms, we may call these the 'ontic commitments' of the theories. A hierarchical ordering of conceptual systems yields a catalogue of ontic commitments ranging from men, through inferior organisms, to fundamental particles or to whatever the physicists of the future shall accept as fundamental: particles, fields, events, or things not yet conceived.

These considerations flow from an analysis of conceptual systems considered as quasi-public objects. Pragmatism enters with the question of which entities should be accepted as basic. Before considering the available options, the epistemological significance of the pragmatic theory of choice should be noted. Ordinary language analysts often hold the position, which Strawson has made explicit, that the descriptive metaphysics of ordinary language is more fundamental than any revision offered by science or philosophy. The basic reason for this is the dependence of the meaning of specialized terms and systems containing such terms on a foundation in ordinary language. In analytic pragmatism basic ontic commitments are determined, not on the grounds of meaning dependence, but by the criterion of which theories or conceptual systems are accepted as the basis for describing and explaining reality. This is similar to the traditional scholastic insistence on judgment as an essential component in knowing, except that the emphasis is shifted from individual propositions to systems, or to external rather than internal questions.

If one follows this approach, the basic problem encountered could be labelled 'reductionism vs. emergence', a contemporary analogue of the earlier mechanist-vitalist controversy. The shared presuppositions that make this confrontation meaningful are: (1) an acceptance of evolution—cosmic, terrestrial, organic, and even social—leading from drifting dust in interstellar space to men exploring the stars; and (2) the *prima facie* operative commitment of higher sciences (using 'higher' in an evolutionary sense) to organisms as irreducible units with distinctive properties and activities. The basic question is whether these commitments are themselves reducible.

A simple, though somewhat artificial, example may clarify the argument's logic. Ordinary arithmetic is functionally committed to numbers as the things one speaks about in arithmetic. If, however, one accepts the Frege-Russell idea that a number may be defined as a class of classes, then the ontic (in a platonic sense) commitment to numbers is reducible to the more basic commitment to classes, because arithmetic is, in principle, reducible to set theory. More generally, if a science, R, is reducible to a more basic science B, then the ontic commitments of R are relatively phenomenological. Such a reductionism need not entail deducing the basic concepts or laws of R from B, as some earlier philosophers supposed. What is required, as a necessary though not sufficient condition, is that it be possible in principle to replace the framework of R by that of B so that there is no explanatory residue (e.g., phenomena that can be explained in R and not in B) and that there is no essential dependence of R on B, though there may be a methodological dependence in fixing the meaning of the concept used in R.

Strict materialistic reductionism would involve the reduction of psychology to neurophysiology, neurophysiology to molecular biology, and so on until all the sciences are considered special cases of fundamental physics. No one argues that such a complete reductionism could be accomplished at the present stage of scientific development. The contention is, rather, that such a reductionism should be possible in principle. The considerations employed to support this contention involve, as a necessary minimum, some demonstration that the proposed program of reducing ontic commitments is self-consistent and non-circular. Even this necessary minimum has not yet been developed in a self-consistent and satisfactory manner.[45] This negative conclusion does not mean that the program has been disproved, but it should induce a certain degree of philosophical restraint in the defenders of reductive materialism. Developing a consistent, critically justifiable materialism is more difficult than is generally supposed.

Emergentists generally do not follow the ground rules outlined above. Thus, Teilhard de Chardin begins by accepting man as a distinctively unique phenomenon and attempts to determine what the past must have been to produce such a present. Nevertheless, an emergentist position should meet the requirements set here as a

45. Reasons supporting this evaluation may be found in my article "The New Materialism," *The Heythrop Journal*, 7 (1967): 5–26.

minimal condition for acceptability. An initial difficulty is the meaning of the term 'emergence'. Much of the disrepute associated with the term is due to explanations of emergence involving either hidden forces in nature (Bergson's *élan vital*), a conscious purposefulness leading evolution to man, or some form of supernatural intervention. The term is also used with regard to entities, properties, laws, or relationships characteristic of higher organisms (or of the appropriate scientific systems) that cannot be explained by reduction to more fundamental units (or to the natural sciences). In accord with the present stress on substantive commitments of theories, we will accept as the basic meaning of the term the emergence of new substantial units that cannot be adequately explained in terms of the properties and activities of their component parts. This is sometimes summarized in the rather misleading statement that the whole is greater than the sum of its parts.

The most defensible development of this doctrine hinges on the idea of organization as a constitutive principle forming an aggregate into a substance or an existential unit. As such it should manifest distinctive properties and characteristic activities not evident in the parts that compose it. Life, sensation, intelligence, and self-determination are generally the basic characteristics used as criteria. That a living being, e.g., a man, is something more than an ordered collection of molecules seems almost self-evident. Unless a program of reductionism can give a plausible explanation of man's unity, consciousness, and capacity for intelligent self-determination, man must be considered an ontologically irreducible unit. Acceptance of this coupled to an acceptance of evolution necessarily entails some doctrine of substantive emergence. However, unless there is a further explanation, e.g., in terms of a metaphysical system, of what it means to be a substance or an existential unit, this minimal doctrine of emergence is hardly distinguishable from many doctrines of reductionism.

A systematic implementation of either program, or of some variant less crude than those sketched above, may lead to a consistent conceptualization of reality compatible with scientific realism. The core of this conceptualization is an acceptance of the entities postulated or presupposed by the substantive scientific theories accepted as irreducible. One who has gone this far is then confronted with a fundamental option. He may decide that such a consistent conceptualization, open-ended enough to admit of further conceptual revolutions, is the best man can hope for and that further metaphysical probing is a quest of a chimera. Such a position embodies the practical realism that we have dubbed 'analytic pragmatism'. Or one may take the critical question seriously and wonder whether this or any other system represents reality as it exists independent of our knowing it. Since we have invoked pragmatic considerations, we should heed the admonition of the founder of pragmatism and not block the path of inquiry by dismissing this critical question without an adequate consideration.

Transcendental Metaphysics

Does the type of conceptualization we have been considering represent the world as it really is? Can at least the main features be considered representations of the really real, of things in themselves? Or do we simply have a constructed system pragmatically justified by its utility in imposing an order on a random collection of stimuli, thus enabling man to adjust to an increasingly complex environment? As Kant saw clearly, any attempt to answer such questions depends upon a critical analysis of man's mode of knowing, its capabilities and limitations. Such a critique is not the subject of the present study. Only one aspect of this question will be considered, the contention that a successful solution of the critical problem yields a metaphysical realism that, in turn, supplies an ultimate basis for interpreting the sort of practical realism or conceptualism so far discussed.

The type of conceptualization already considered could be labeled 'ontology within a naturalistic framework'. Husserl and later phenomenologists have strongly insisted that such an ontology is radically inadequate as a basis for a critically developed metaphysics. I believe that this negative appraisal is essentially correct, but that the phenomenologists' insistence on this point has led them to ignore the function of such an ontology and to misjudge the basic role it should play in the development of an adequate philosophy. The relationship that should, in my opinion, obtain can be approached by considering how one might pass from the naturalistic ontology, or consistent conceptualization considered above, to a metaphysics of being. We shall mention three such methods. Although mentioned separately, they should, I believe, be used in conjunction.

Transcendental deduction If one accepts systems of knowledge of the sort considered earlier as a relatively adequate expression of man's knowledge, then one might ask what must the world be like if it can be known in the way we do in fact know it? [46] This approach may set boundary values on admissible answers, but is inadequate unless one considers, not only the expression of knowledge through systems, but also the dynamics of knowing. This leads to:

The dialectic of knowing The stress on conceptual analysis as well as the particular way in which the critical question was phrased may give the impression that one need only concern oneself with idealized conceptual systems, much the way Kant did with his categories or the early Wittgenstein did with the ideal language of the *Tractatus*. But knowledge does not function in this way, for both the community and the individual develop intellectually through an interaction with the world. Thus, one comes to understand a puzzling situation by considering the different possibilities that could obtain and discovering which of these actually does. In the development of knowledge man proposes through hypothesis and systems, but eventually nature disposes. Accordingly, one must consider, not merely the content of the known, but also the interaction with reality involved in acquiring, developing and testing knowledge. This, of course, requires a cognitional analysis similar to that initiated by Lonergan. But the focus of such an analysis, for our present purposes, is directed toward nature's contribution to the on-going man-nature dialogue. This leads to a consideration of:

Getting to the pre-conceptual given One may almost define naive realism in terms of adherence to the myth of the given, to use Sellars's apt phrase. The myth, or at least one of its variant forms, teaches that basic concepts of ordinary language or of some approved philosophical system represent reality because they are

46. Though Körner denies the validity of a transcendental deduction in a technical sense, his development illustrates the limited sort of transcendental deduction intended here. This is discussed further in my article "Epistemological Problems n the Philosophy of Science."

somehow either given in immediate experience or derived by a quasi-automatic abstraction from experience. The reaction against this epistemological naiveté sometimes induces the opposite extreme of minimizing the aspect of givenness in knowledge and confining it to the limits prescribed by a nominalistic epistemology. Neither epistemological naiveté nor system-centered apriorism supplies a satisfactory means of coming to grips with this problem.

If knowledge of external reality necessarily involves something given from without, as well as something imposed from within, and if the given cannot be determined by isolating key concepts that are thought to represent a transcription of the given elements into language or thought, then a more adequate means of analysis must be found. Husserl, Heidegger, Merleau-Ponty, and others have attempted to isolate and express this given element, or at least the conditions for the reception of such elements by means of a protracted phenomenological analysis. These analyses are not of direct concern here since they did not serve to ground the type of metaphysical realism now being discussed. However, the transcendental Thomists, considered earlier, claim to have incorporated the basic insights and methods of these phenomenologists and to have developed a rigorous and permanently valid metaphyiscal realism.

In our earlier consideration of transcendental Thomism, we listed two essential shortcomings in these developments. First, the scientific world view was excluded as essentially irrelevant since the metaphysics attempts to build exclusively on the elements attained by transcendental reduction. Secondly, the crucial role of thematization is effectively ignored, so that the metaphysics actually builds, not on the transcendental *a priori* conditions of knowing, but on their thematization in language. This means that the proponents of this methodology are doing one of two things. Either they are backing into a doctrine of private language, if they think they can express the pre-conceptual given while bracketing the shared conceptualization of reality implicit in any expression of this given, or they are resorting to the quasi-mystical doctrine of an *Urlogos* in nature somehow grounding the meaningfulness of language, a fundamental theme in the later writings of Heidegger.[47]

47. See W. J. Richardson, S.J., "Heidegger and the Origin of Language," *International Philosophical Quarterly,* 2 (1962): 404–16.

We would like to suggest a possible way of redeveloping what is valid in this tradition while avoiding these shortcomings. The thematization in question is the expression in meaningful language of the transcendental *a priori* conditions of the possibility of the act of knowing. Here one must distinguish between getting at these elements by phenomenological analysis and expressing the conclusions of this analysis. Since this expression plays a crucial role, one can treat it critically only if he has already come to realize what is involved in expressing anything.

Language is a symbolic vehicle; it is, in fact, the primary symbolic vehicle. Using language to express a theory means constructing in this symbolic vehicle a conceptual analogue intended to represent structural features and essential relations of a certain domain of reality. Language necessarily objectifies even when speaking of subjectivity. In order for the phenomenologist to discuss the preconceptual given of transcendental subjectivity or whatever it is he wishes to discuss, he must construct a theory about the aspects of reality being considered. As an expressed theory it is not essentially different from other theories, though the grounds of accepting, testing, evaluating, and confirming it are different. The role of theorizing is most easily seen and most critically analyzed by a prior study of more objective theories, e.g., the type of theories we have been considering. With such a prior analysis one can avoid such errors as claiming that the crucial terms employed have meaning by virtue of a relation to the pre-conceptual given that the theory seeks to express. There is an essential difference between trying to get at the pre-conceptual given through an analysis of subjectivity or through some form of introspection, such as Longergan's cognitional analysis and, trying to express what one might have discovered.

This explicit recognition of the role of theories allows for a linking of metaphysics, so developed, and the pre-critical ontology developed within a naturalistic framework, provided certain conditions are met. From the naturalistic ontology one must accept some doctrine of emergence with the conclusion that man, as well as other organisms, is an existential unit, though this ontology does not explain what it means to be an existential unit. From a phenomenological analysis of subjectivity, of man as the *Dasein* of being, in the privileged case of self-affirmation, one can know what it means to know being as being and can express this understanding in meaningful language.

Trying to understand other entities as beings means, in this development, trying to understand them by analogy with our understanding of man affirmed as a being. This means that we employ the conclusions of the pre-critical metaphysics with its ontic commitment to entities ranging from fundamental particles to mammals, but ask different questions about these entities. In the naturalistic ontology, we attempted to understand from the bottom up the grades of being that resulted from evolutionary emergence, in accord with the materialism methodologically presupposed. Now we are seeking to understand the same entities from the top down, that is, to take an idea of being and its constitutive principles derived from the analysis of man considered as being and to extend these principles to other beings. The result of successfully extending and elaborating these principles would be a concrete existential metaphysics.

The significance of these suggestions can best be seen by summarizing the conclusions drawn in this section. Science as practiced has a functional commitment to realism. The basis of this commitment is the realization that the scientist can name entities and make statements about these entities that are accepted as true in a correspondence sense of 'true'. Such ontic commitments are, I have argued, essentially practical, revisable, and pre-critical. They are the core of an interrelated set of conceptual systems that express a scientific understanding of reality, but do not, in themselves, answer specifically philosophical questions.

There are also various types of philosophical realism. The question that concerned us was whether any of these systems could supply either a critical vindication of some sort of scientific realism or an adequate basis for evaluating its validity. Immediate realism was examined and eventually rejected as not having an adequate critical justification. Analytic pragmatism essentially transforms the critical question from its speculative form, "Can we know things as they are in themselves?" into the practical question, "Which classification and description of things is the most useful to accept?" This is a reasonable practical position, but need be accepted as the terminus of the philosophical quest only if one has either rejected the critical question as meaningless or unanswerable or has concluded that a metaphysics of being is impossible.

'Transcendental metaphysics' was used as a general label referring to any development of metaphysics that accepts the critical question, usually in a neo-Kantian form, as a necessary prolegomenon to the development of metaphysics. Here I concentrated on one particular school, the transcendental Thomists. While my metaphysical sympathies lie with this group, these sympathies are not strong enough to blot out critical objections. The suggestions I offered for avoiding the basic difficulties in this approach are underdeveloped and as yet untested. Accordingly, I must conclude that a truly adequate, critically developed metaphysical underpinning for scientific realism is not available at present. Science can certainly survive and even flourish without such an underpinning. Yet, as Kant saw so clearly, the attempt to ask and answer such ultimate questions flows from the thrust of human knowing. The search must and will go on.

READINGS

Aristotle

Aristotle's systematization of science and scientific explanation dominated the development of science for nearly two thousand years. The following selections bring out both his ideas on the nature of scientific explanation and the methodology he followed in developing particular sciences. The first excerpt, from Book I of his *Posterior Analytics,* clarifies the nature of deductive explanation and summarizes the necessary qualities of the premises of a deductive argument. The second excerpt, from Book II of the *Posterior Analytics,* deals with the reasoning leading from these premises to conclusions, particularly the relation between middle terms and causes. The selection from the *Physics* illustrates Aristotle's operative methodology. After clarifying the presuppositions and confused common knowledge that is the starting point for inquiry, he begins an analysis aimed at uncovering the principles that will serve to explain the changes material bodies undergo. The final brief excerpt from the *Metaphysics* summarizes Aristotle's ideas on the interrelation of physics, mathematics, and metaphysics.

Science as a Systematic Explanation Through Causes

ANALYTICA POSTERIORA (POSTERIOR ANALYTICS)

Book 1

1

All instruction given or received by way of argument proceeds from pre-existent knowledge. This becomes evident upon a survey of all the species of such instruction. The mathematical sciences and all other speculative disciplines are acquired in this way, and so are the two forms of dialectical reasoning, syllogistic and inductive; for each of these latter makes use of old knowledge to impart new, the syllogism assuming an audience that accepts its premises, induction exhibiting the universal as implicit in the clearly known particular. Again, the persuasion exerted by rhetorical arguments is in principle the same, since they use either example, a kind of induction, or enthymeme, a form of syllogism.

The pre-existent knowledge required is of two kinds. In some cases admission of the fact must be assumed, in others comprehension of the meaning of the term used, and sometimes both assumptions are essential. Thus, we assume that every predicate can be either truly affirmed or truly denied of any subject, and that

These selections are reprinted from the *Oxford Translation of Aristotle,* by permission of the Clarendon Press, Oxford.

"triangle" means so and so; as regards "unit" we have to make the double assumption of the meaning of the word and the existence of the thing. The reason is that these several objects are not equally obvious to us. Recognition of a truth may in some cases contain as factor both previous knowledge and also knowledge acquired simultaneously with that recognition—knowledge, this latter, of the particulars actually falling under the universal and therein already virtually known. For example, the student knew beforehand that the angles of every triangle are equal of two right angles; but it was only at the actual moment at which he was being led on to recognize this as true in the instance before him that he came to know "this figure inscribed in the semicircle" to be a triangle. For some things (viz. the singulars finally reached which are not predictable of anything else as subject) are only learnt in this way, i.e., there is here no recognition through a middle of a minor term as subject to a major. Before he was led on to recognition or before he actually drew a conclusion, we should perhaps say that in a manner he knew, in a manner not.

If he did not in an unqualified sense of the term *know* the existence of this triangle, how could he *know* without qualification that its angles were equal to two right angles? No: clearly he *knows* not without qualification but only in the sense that he *knows* universally. If this distinction is not drawn, we are faced with the dilemma in the *Meno:* [1] either a man will learn nothing or what he already knows; for we cannot accept the solution which some people offer. A man is asked, "Do you, or do you not, know that every pair is even?" He says he does know it. The questioner then produces a particular pair, of the existence, and so *a fortiori* of the evenness, of which he was unaware. The solution which some people offer is to assert that they do not know that every pair is even, but only that everything which they know to be a pair is even: yet what they know to be even is that of which they have demonstrated evenness, i.e. what they made the subject of their premiss, viz. not merely every triangle or number which they know to be such, but any and every number or triangle without reservation. For no premiss is ever couched in the form "every number which you know to be such," or "every rectilinear figure which you know to be such":

1. Plato, *Meno,* 80 E.

the predicate is always construed as applicable to any and every instance of the thing. On the other hand, I imagine there is nothing to prevent a man in one sense knowing what he is learning, in another not knowing it. The strange thing would be, not if in some sense he knew what he was learning, but if he were to know it in that precise sense and manner in which he was learning it.[2]

2

We suppose ourselves to possess unqualified scientific knowledge of a thing, as opposed to knowing it in the accidental way in which the sophist knows, when we think that we know the cause on which the fact depends, as the cause of that fact and of no other, and, further, that the fact could not be other than it is. Now that scientific knowing is something of this sort is evident—witness both those who falsely claim it and those who actually possess it, since the former merely imagine themselves to be, while the latter are also actually, in the condition described. Consequently the proper object of unqualified scientific knowledge is something which cannot be other than it is.

There may be another manner of knowing as well—that will be discussed later.[3] What I now assert is that at all events we do know by demonstration. By demonstration I mean a syllogism productive of scientific knowledge, a syllogism, that is, the grasp of which is *eo ipso* such knowledge. Assuming then that my thesis as to the nature of scientific knowing is correct, the premisses of demonstrated knowledge must be true, primary, immediate, better known than and prior to the conclusion, which is further related to them as effect to cause. Unless these conditions are satisfied, the basic truths will not be "appropriate" to the conclusion. Syllogism there may indeed be without these conditions, but such syllogism, not being productive of scientific knowledge, will not be demonstration. The premisses must be true: for that which is non-existent cannot be known—we cannot know, e.g., that the diagonal of a square is commensurate with its side. The premisses must be primary and indemonstrable; otherwise they will require demonstration in order to be known, since to have knowledge, if it be not

2. Cf. *An. Pr.* ii, ch. 21.
3. Cf. *i,* ch. 2 and more particularly ii, ch. 19.

accidental knowledge, of things which are demonstrable, means precisely to have a demonstration of them. The premisses must be the causes of the conclusion, better known than it, and prior to it; its causes, since we possess scientific knowledge of a thing only when we know its cause; prior, in order to be causes; antecedently known, this antecedent knowledge being not our mere understanding of the meaning, but knowledge of the fact as well. Now "prior" and "better known" are ambiguous terms, for there is a difference between what is prior and better known in the order of being and what is prior and better known to man. I mean that objects nearer to sense are prior and better known to man; objects without qualification prior and better known are those further from sense. Now the most universal causes are furthest from sense and particular causes are nearest to sense, and they are thus exactly opposed to one another. In saying that the premisses of demonstrated knowledge must be primary, I mean that they must be the "appropriate" basic truths, for I identify primary premiss and basic truth. A 'basic truth' in a demonstration is an immediate proposition. An immediate proposition is one which has no other proposition prior to it. A proposition is either part of an enunciation, i.e. it predicates a single attribute of a single subject. If a proposition is dialectical, it assumes either part indifferently; if it is demonstrative, it lays down one part to the definite exclusion of the other because that part is true. The term 'enunciation' denotes either part of a contradiction indifferently. A contradiction is an opposition which of its own nature excludes a middle. The part of a contradiction which conjoins a predicate with a subject is an affirmation; the part disjoining them is a negation. I call an immediate basic truth of syllogism a 'thesis' when, though it is not susceptible of proof by the teacher, yet ignorance of it does not constitute a total bar to progress on the part of the pupil: one which the pupil must know if he is to learn anything whatever is an axiom. I call it an axiom because there are such truths and we give them the name of axioms *par excellence*. If a thesis assumes one part or the other of an enunciation, i.e. asserts either the existence or the non-existence of a subject, it is a hypothesis; if it does not so assert, it is a definition. Definition *is* a 'thesis' or a "laying something down," since the arithmetician lays it down that to be a unit is to be quantitatively indivisible; but it is not a hypothesis, for to define what a unit is is not the same as to affirm its existence.

Now since the required ground of our knowledge—i.e. of our conviction—of a fact is the possession of such a syllogism as we call demonstration, and the ground of the syllogism is the facts constituting its premisses, we must not only know the primary premisses—some if not all of them—beforehand, but know them better than the conclusion: for the cause of an attribute's inherence in a subject always itself inheres in the subject more firmly than that attribute; e.g. the cause of our loving anything is dearer to us than the object of our love. So since the primary premisses are the cause of our knowledge—i.e. of our conviction—it follows that we know them better—that is, are more convinced of them—than their consequences, precisely because our knowledge of the latter is the effect of our knowledge of the premisses. Now a man cannot believe in anything more than in the things he knows, unless he has either actual knowledge of it or something better than actual knowledge. But we are faced with this paradox if a student whose belief rests on demonstration has not prior knowledge; a man must believe in some, if not in all, of the basic truths more than in the conclusion. Moreover, if a man sets out to acquire the scientific knowledge that comes through demonstration, he must not only have a better knowledge of the basic truths and a firmer conviction of them than of the connexion which is being demonstrated: more than this, nothing must be more certain or better known to him than these basic truths in their character as contradicting the fundamental premisses which lead to the opposed and erroneous conclusion. For indeed the conviction of pure science must be unshakable.

3

Some hold that, owing to the necessity of knowing the primary premisses, there is no scientific knowledge. Others think there is, but that all truths are demonstrable. Neither doctrine is either true or a necessary deduction from the premisses. The first school, assuming that there is no way of knowing other than by demonstration, maintain that an infinite regress is involved, on the ground that if behind the prior stands no primary, we could not know the posterior through the prior (wherein they are right, for one cannot traverse an infinite series): if on the other hand—they say—the series terminates and there are primary premisses, yet these are unknowable because incapable of demonstration, which according to them is the only form of knowledge. And since thus one cannot know the primary premisses, knowledge of the conclusions which follow from them is not pure scientific knowledge nor properly knowing at all, but rests on the mere supposition that the premisses are true. The other party agrees with them as regards knowing, holding that it is only possible by demonstration, but they see no difficulty in holding that all truths are demonstrated, on the ground that demonstration may be circular and reciprocal.

Our own doctrine is that not all knowledge is demonstrative: on the contrary, knowledge of the immediate premisses is independent of demonstration. (The necessity of this is obvious; for since we must know the prior premisses from which the demonstration is drawn, and since the regress must end in immediate truths, those truths must be indemonstrable.) Such, then, is our doctrine, and in addition we maintain that besides scientific knowledge there is its originative source which enables us to recognize the definitions.

Now demonstration must be based on premisses prior to and better known than the conclusion; and the same things cannot simultaneously be both prior and posterior to one another: so circular demonstration is clearly not possible in the unqualified sense of 'demonstration,' but only possible if 'demonstration' be extended to include that other method of argument which rests on a distinction between truths prior to us and truths without qualification prior, i.e. the method by which induction produces knowledge. But if we accept this extension of its meaning, our definition of unqualified knowledge will prove faulty; for there seem to be two kinds of it. Perhaps, however, the second form of demonstration, that which proceeds from truths better known to us, is not demonstration in the unqualified sense of the term.

The advocates of circular demonstration are not only faced with the difficulty we have just stated: in addition their theory reduces to the mere statement that if a thing exists, then it does exist—an easy way of proving anything. That this is so can be clearly shown by taking three terms, for to constitute the circle it makes no difference whether many terms or few or even only two are taken. Thus by direct proof, if A is, B must be; if B is, C must be; therefore if A is, C must be. Since then—by the circular proof—if A is, B must be, and if B is, A must be, A may be substituted for C above. Then "if B is, A must be" = "if B is, C must be," which above gave the conclusion "if *A* is, *C* must be": but *C* and *A* have been identified. Consequently the upholders of circular demonstration are in the position of saying that if *A* is, *A* must be—a simple way of proving anything. Moreover, even such circular demonstration is impossible except in the case of attributes that imply one another, viz. "peculiar" properties.

Now, it has been shown that the positing of one thing—be it one term or one premiss—never involves a necessary consequent: [4] two premisses constitute the first and smallest foundation for drawing a conclusion at all and therefore *a fortiori* for the demonstrative syllogism of science. If, then, *A* is implied in *B* and *C*, and *B* and *C* are reciprocally implied in one another and in *A*, it is possible, as has been shown in my writings on the syllogism,[5] to prove all the assumptions on which the original conclusion rested, by circular demonstration in the first figure. But it has also been shown that in other figures either no conclusion is possible, or at least none which proves both the original premisses.[6] Propositions the terms of which are not convertible cannot be circularly demonstrated at all, and since convertible terms occur rarely in actual demonstrations, it is clearly frivolous and impossible to say that demonstration is reciprocal and that therefore everything can be demonstrated.

Book II

1

The kinds of question we ask are as many as the kinds of things which we know. They are in fact four:—(1) whether the connexion of an attribute with a thing is a fact, (2) what is the reason of the connexion, (3) whether a thing exists, (4) what is the nature of the thing. Thus, when our question concerns a complex of thing and attribute and we ask whether the thing is thus or otherwise qualified—whether, e.g., the sun suffers eclipse or not—then we are asking as to the fact of a connexion. That our inquiry ceases with the discovery that the sun does suffer eclipse is an indication

4. *An. Pr.* i, ch. 25.
5. *Ibid.* ii, ch. 5.
6. *Ibid.* ii, cc. 5 and 6.

of this; and if we know from the start that the sun suffers eclipse, we do not inquire whether it does so or not. On the other hand, when we know the fact we ask the reason; as, for example, when we know that the sun is being eclipsed and that an earthquake is in progress, it is the reason of eclipse or earthquake into which we inquire.

Where a complex is concerned, then, those are the two questions we ask; but for some objects of inquiry we have a different kind of question to ask, such as whether there is or is not a centaur or a God. (By "is or is not" I mean "is or is not, without further qualification"; as opposed to "is or is not (e.g.) white.") On the other hand, when we have ascertained the thing's existence, we inquire as to its nature, asking, for instance, "what, then, is God?" or "what is man?"

2

These, then, are the four kinds of question we ask, and it is in the answers to these questions that our knowledge consists.

Now when we ask whether a connexion is a fact, or whether a thing without qualification *is,* we are really asking whether the connexion or the thing has a 'middle'; and when we have ascertained either that the connexion is a fact or that the thing *is*—i.e. ascertained either the partial or the unqualified being of the thing—and are proceeding to ask the reason of the connexion or the nature of the thing, then we are asking what the 'middle' is.

(By distinguishing the fact of the connexion and the existence of the thing as respectively the partial and the unqualified being of the thing, I mean that if we ask "does the moon suffer eclipse?" or "does the moon wax?" the question concerns a part of the thing's being; for what we are asking in such questions is whether a thing is this or that, i.e. has or has not this or that attribute: whereas, if we ask whether the moon or night exists, the question concerns the unqualified being of a thing.)

We conclude that in all our inquiries we are asking either whether there is a 'middle' or what the 'middle' is: for the 'middle' here is precisely the cause, and it is the cause that we seek in all our inquiries. Thus, "Does the moon suffer eclipse?" means "Is there or is there not a cause producing eclipse of the moon?" and when we have learnt that there is, our next question is, "What, then, is this cause?"; for the cause through which a thing *is*—not *is this or that,* i.e. has this or that attribute, but without qualification *is*— and the cause through which it is—not *is* without qualification, but *is this or that* as having some essential attribute or some accident—are both alike the 'middle'. By that which *is* without qualification I mean the subject, e.g. moon or earth or sun or triangle; by that which a subject *is* (in the partial sense) I mean a property, e.g. eclipse, equality or inequality, interposition or non-interposition. For in all these examples it is clear that the nature of the thing and the reason of the fact are identical: the question "What is eclipse?" and its answer "The privation of the moon's light by the interposition of the earth" are identical with the question "What is the reason of eclipse?" or "Why does the moon suffer eclipse?" and the reply "Because of the failure of light through the earth's shutting it out." Again, for "What is a concord? A commensurate numerical ratio of a high and a low note," we may substitute "What reason makes a high and a low note concordant? Their relation according to a commensurate numerical ratio." "Are the high and the low note concordant?" is equivalent to "Is their ratio commensurate?"; and when we find that it is commensurate, we ask "What, then, is their ratio?"

Cases in which the 'middle' is sensible show that the object of our inquiry is always the 'middle': we inquire, because we have not perceived it, whether there is or is not a 'middle' causing e.g. an eclipse. On the other hand, if we were on the moon we should not be inquiring either as to the fact or the reason, but both fact and reason would be obvious simultaneously. For the act of perception would have enabled us to know the universal too; since, the present fact of an eclipse being evident, perception would then at the same time give us the present fact of the earth's screening the sun's light, and from this would arise the universal.

Thus, as we maintain, to know a thing's nature is to know the reason why it is; and this is equally true of things in so far as they are said without qualification to *be* as opposed to being possessed of some attribute, and in so far as they are said to be possessed of some attribute such as equal to two right angles, or greater or less.

3

It is clear, then, that all questions are a search for a 'middle'. Let us now state how essential nature is revealed, and in what way it can be reduced to demonstration; [7] what definition is, and what things are definable. And let us first discuss certain difficulties which these questions raise, beginning what we have to say with a point most intimately connected with our immediately preceding remarks, namely the doubt that might be felt as to whether or not it is possible to know the same thing in the same relation, both by definition and by demonstration. It might, I mean, be urged that definition is held to concern essential nature and is in every case universal and affirmative; whereas, on the other hand, some conclusions are negative and some are not universal; e.g., all in the second figure are negative, none in the third are universal. And again, not even all affirmative conclusions in the first figure are definable, e.g., "every triangle has its angles equal to two right angles." An argument proving this difference between demonstration and definition is that to have scientific knowledge of the demonstrable is identical with possessing a demonstration of it; hence if demonstration of such conclusions as these is possible, there clearly cannot also be definition of them. If there could, one might know such a conclusion also in virtue of its definition without possessing the demonstration of it; for there is nothing to stop our having the one without the other.

Induction too will sufficiently convince us of this difference; for never yet by defining anything—essential attribute or accident—did we get knowledge of it. Again, if to define is to acquire knowledge of a substance, at any rate such attributes are not substances.

It is evident, then, that not everything demonstrable can be defined. What then? Can everything definable be demonstrated, or not? There is one of our previous arguments which covers this too. Of a single thing *qua* single there is a single scientific knowledge. Hence, since to know the demonstrable scientifically is to possess the demonstration of it, an impossible consequence will follow:—possession of its definition without its demonstration will give knowledge of the demonstrable.

7. Cf. 94ª 11–14.

Moreover, the basic premisses of demonstrations are definitions, and it has already been shown [8] that these will be found indemonstrable; either the basic premisses will be demonstrable and will depend on prior premisses, and the regress will be endless; or the primary truths will be indemonstrable definitions.

But if the definable and the demonstrable are not wholly the same, may they yet be partially the same? Or is that impossible, because there can be no demonstration of the definable? There can be none, because definition is of the essential nature or being of something, and all demonstrations evidently posit and assume the essential nature—mathematical demonstrations, for example, the nature of unity and the odd, and all the other sciences likewise. Moreover, every demonstration proves a predicate of a subject as attaching or as not attaching to it, but in definition one thing is not predicated of another; we do not, e.g., predicate animal of biped nor biped of animal, nor yet figure of plane—plane not being figure nor figure plane. Again, to prove essential nature is not the same as to prove the fact of a connexion. Now definition reveals essential nature, demonstration reveals that a given attribute attaches or does not attach to a given subject; but different things require different demonstrations—unless the one demonstration is related to the other as part to whole. I add this because if all triangles have been proved to possess angles equal to two right angles, then this attribute has been proved to attach to isosceles; for isosceles is a part of which all triangles constitute the whole. But in the case before us the fact and the essential nature are not so related to one another, since the one is not a part of the other.

So it emerges that not all the definable is demonstrable nor all the demonstrable definable; and we may draw the general conclusion that there is no identical object of which it is possible to possess both a definition and a demonstration. It follows obviously that definition and demonstration are neither identical nor contained either within the other: if they were, their objects would be related either as identical or as whole and part.

8. Cf. 72[b] 18–25 and 84[a] 30–[b] 2.

PHYSICA (PHYSICS)

Book I

1

When the objects of an inquiry, in any department, have principles, conditions, or elements, it is through acquaintance with these that knowledge, that is to say scientific knowledge, is attained. For we do not think that we know a thing until we are acquainted with its primary conditions or first principles, and have carried our analysis as far as its simplest elements. Plainly therefore in the science of Nature, as in other branches of study, our first task will be to try to determine what relates to its principles.

The natural way of doing this is to start from the things which are more knowable and obvious to us and proceed towards those which are clearer and more knowable by nature; for the same things are not "knowable relatively to us" and "knowable" without qualification. So in the present inquiry we must follow this method and advance from what is more obscure by nature, but clearer to us, towards what is more clear and more knowable by nature.

Now what is to us plain and obvious at first is rather confused masses, the elements and principles of which become known to us later by analysis. Thus we must advance from generalities to particulars; for it is a whole that is best known to sense-perception, and a generality is a kind of whole, comprehending many things within it, like parts. Much the same thing happens in the relation of the name to the formula. A name, e.g., "round," means vaguely a sort of whole: its definition analyses this into its particular senses. Similarly a child begins by calling all men "father," and all women "mother," but later on distinguishes each of them.

The present treatise, usually called the *Physics*, deals with natural body in general: the special kinds are discussed in Aristotle's other physical works, the *De Caelo*, &c. The first book is concerned with the elements of a natural body (matter and form): the second mainly with the different types of cause studied by the physicist. Books III–VII deal with movement, and the notions implied in it. The subject of VIII is the prime mover, which, though not itself a natural body, is the cause of movement in natural bodies.

2

The principles in question must be either (a) one or (b) more than one.

If (a) one, it must be either (i) motionless, as Parmenides and Melissus assert, or (ii) in motion, as the physicists hold, some declaring air to be the first principle, others water.

If (b) more than one, then either (i) a finite or (ii) an infinite plurality. If (i) finite (but more than one), then either two or three or four or some other number. If (ii) infinite, then either as Democritus believed one in kind, but differing in shape or form; or different in kind and even contrary.

A similar inquiry is made by those who inquire into the number of *existents*: for they inquire whether the ultimate constituents of existing things are one or many, and if many, whether a finite or an infinite plurality. So they too are inquiring whether the principle or element is one or many.

Now to investigate whether Being is one and motionless is not a contribution to the science of Nature. For just as the geometer has nothing more to say to one who denies the principles of his science—this being a question for a different science or for one common to all—so a man investigating *principles* cannot argue with one who denies their existence. For if Being is just one, and one in the way mentioned, there is a principle no longer, since a principle must be the principle of some thing or things.

To inquire therefore whether Being is one in this sense would be like arguing against any other position maintained for the sake of argument (such as the Heraclitean thesis, or such a thesis as that Being is one man) or like refuting a merely contentious argument —a description which applies to the arguments both of Melissus and of Parmenides: their premises are false and their conclusions do not follow. Or rather the argument of Melissus is gross and palpable and offers no difficulty at all: accept one ridiculous proposition and the rest follows—a simple enough proceeding.

We physicists, on the other hand, must take for granted that the things that exist by nature are, either all or some of them, in motion—which is indeed made plain by induction. Moreover, no man of science is bound to solve the kind of difficulty that may be raised, but only as many as are drawn falsely from the principles of the science: it is not our business to refute those that do not arise in this way: just as it is the duty of the geometer to refute the squaring of the circle by means of segments, but it is not his duty to refute Antiphon's proof. At the same time the holders of the theory of which we are speaking do incidentally raise physical questions, though Nature is not their subject: so it will perhaps be as well to spend a few words on them, especially as the inquiry is not without scientific interest.

Book E (VI)

1

We are seeking the principles and the causes of the things that are, and obviously of them *qua* being. For, while there is a cause of health and of good condition, and the objects of mathematics have first principles and elements and causes, and in general every science which is ratiocinative or at all involves reasoning deals with causes and principles, more or less precise, all these sciences mark off some particular being—some genus, and inquire into this, but not into being simply nor *qua* being, nor do they offer any discussion of the essence of the things of which they treat; but starting from the essence—some making it plain to the senses, others assuming it as a hypothesis—they then demonstrate, more or less cogently, the essential attributes of the genus with which they deal. It is obvious, therefore, that such an induction yields no demonstration of substance or of the essence, but some other way of exhibiting. And similarly the sciences omit the question whether the genus with which they deal exists or does not exist, because it belongs to the same kind of thinking to show what it is and that it is.

And since natural science, like other sciences, is in fact about one class of being, i.e. to that sort of substance which has the principle of its movement and rest present in itself, evidently it is neither practical nor productive. For in the case of things made the principle is in the maker—it is either reason or art or some faculty, while in the case of things done it is in the doer—viz. will, for that which is done and that which is willed are the same. Therefore, if all thought is either practical or productive or theoretical, physics must be a theoretical science, but it will theorize about such being as admits of being moved, and about substance-as-defined for the most part only as not separable from matter. Now, we must not fail to notice the mode of being of the essence and of its definition, for, without this, inquiry is but idle. Of things defined, i.e. of "whats," some are like "snub," and some like "concave." And these differ because "snub" is bound up with matter (for what is snub is a concave *nose*), while concavity is independent of perceptible matter. If then all natural things are analogous to the snub in their nature—e.g. nose, eye, face, flesh, bone, and, in general, animal; leaf, root, bark, and, in general, plant (for none of these can be defined without reference to movement—they always have matter), it is clear how we must seek and define the "what" in the case of natural objects, and also that it belongs to the student of nature to study even soul in a certain sense, i.e. so much of it as is not independent of matter.

That physics, then, is a theoretical science, is plain from these considerations. Mathematics also, however, is theoretical; but whether its objects are immovable and separable from matter, is not at present clear; still, it is clear that *some* mathematical theorems *consider* them *qua* immovable and *qua* separable from matter. But if there is something which is eternal and immovable and separable, clearly the knowledge of it belongs to a theoretical science—not, however, to physics (for physics deals with certain movable things) nor to mathematics, but to a science prior to both. For physics deals with things which exist separately but are not immovable, and some parts of mathematics deal with things which are immovable but presumably do not exist separately, but as embodied in matter; while the first science deals with things which both exist separately and are immovable. Now all causes must be eternal, but especially these; for they are the causes that operate on so much of the divine as appears to us.[1] There must, then, be three theoretical philosophies, mathematics, physics, and what we may call theology, since it is obvious that if the divine is present anywhere, it is present in things of this sort. And the highest science must deal with the highest genus. Thus, while the theoretical sci-

---

1. i.e. produce the movements of the heavenly bodies.

ences are more to be desired than the other sciences, this is more to be desired than the other theoretical sciences. For one might raise the question whether first philosophy is universal, or deals with one genus, i.e. some one kind of being; for not even the mathematical sciences are all alike in this respect—geometry and astronomy deal with a certain particular kind of thing, while universal mathematics applies alike to all. We answer that if there is no substance other than those which are formed by nature, natural science will be the first science; but if there is an immovable substance, the science of this must be prior and must be first philosophy, and universal in this way, because it is first. And it will belong to this to consider being *qua* being—both what it is and the attributes which belong to it *qua* being.[2]

2. With ch. I Cf. iii. 995^b 10–13, 997^a 15–25, xi. 7.

Isaac Newton

Isaac Newton (1642–1727) is generally recognized as the most influential figure in the history of physics. Although he never wrote a systematic treatise on scientific explanation, he occasionally felt constrained to explain what he was doing, why, and how. These philosophical asides were generally intended to clear up misunderstandings stemming from Aristotelian or Cartesian ideas on the nature of scientific explanation or to answer objections brought against his own views. In Books I and II of his monumental *Mathematical Principles of Natural Philosophy,* Newton develops the mathematical principles, while in Book III he applies these principles to observable bodies. The rules of reasoning, from the beginning of Book III, present a very inductive view of scientific explanation. This exposition undoubtedly represents an over-reaction against the Cartesian doctrine that scientific explanation is based, not on observation, but on logical deduction from principles known by pure intuition. Newton appended a *General Scholion* to the second edition of the *Principia* to refute some theological and philosophical objections against his views. The excerpt reproduced here shows his distinctive break with the Aristotelian idea that science is an explanation through causes. In spite of Newton's strictures against the use of hypotheses in science, he did introduce speculative hypotheses. In the "Queries" appended to his *Opticks,* in particular, Newton speculates on issues that were beyond the reach of the science of his day. The excerpt included here not only brings out his speculative views on the nature of matter and the methodology of scientific explanation, but it also illustrates the peculiar way in which Newton read divine design into his theories of the universe.

Rules and Reflections on Scientific Reasoning

RULES OF REASONING IN PHILOSOPHY

Rule I

We are to admit no more causes of natural things than such as are true and sufficient to explain their appearances.

To this purpose the philosophers say that Nature does nothing in vain, and more is in vain when less will serve; for Nature is pleased with simplicity, and affects not the pomp of superfluous causes.

Rule II

Therefore to the same natural effects we must, as far as possible, assign the same causes.

As to respiration in a man and in a beast; the descent of stones in *Europe* and in *America;* the light of our culinary fire and of the sun; the reflection of light in the earth, and in the planets.

Rule III

The qualities of bodies, which admit neither intensification nor remission of degrees, and which are found to belong to all bodies within the reach of our experiments, are to be esteemed the universal qualities of all bodies whatsoever.

From Sir Isaac Newton, *Principia Mathematica,* pp. 398–400, 546–47, by permission of the University of California Press.

For since the qualities of bodies are only known to us by experiments, we are to hold for universal all such as universally agree with experiments; and such as are not liable to diminution can never be quite taken away. We are certainly not to relinquish the evidence of experiments for the sake of dreams and vain fictions of our own devising; nor are we to recede from the analogy of Nature, which is wont to be simple, and always consonant to itself. We no other way know the extension of bodies than by our senses, nor do these reach it in all bodies; but because we perceive extension in all that are sensible, therefore we ascribe it universally to all others also. That abundance of bodies are hard, we learn by experience; and because the hardness of the whole arises from the hardness of the parts, we therefore justly infer the hardness of the undivided particles not only of the bodies we feel but of all others. That all bodies are impenetrable, we gather not from reason, but from sensation. The bodies which we handle we find impenetrable, and thence conclude impenetrability to be an universal property of all bodies whatsoever. That all bodies are movable, and endowed with certain powers (which we call the inertia) of persevering in their motion, or in their rest, we only infer from the like properties observed in the bodies which we have seen. The extension, hardness, impenetrability, mobility, and inertia of the whole, result from the extension, hardness, impenetrability, mobility, and inertia of the parts; and hence we conclude the least particles of all bodies to be also all extended, and hard and impenetrable, and movable, and endowed with their proper inertia. And this is the foundation of all philosophy. Moreover, that the divided but contiguous particles of bodies may be separated from one another, is matter of observation; and, in the particles that remain undivided, our minds are able to distinguish yet lesser parts, as is mathematically demonstrated. But whether the parts so distinguished, and not yet divided, may, by the powers of Nature, be actually divided and separated from one another, we cannot certainly determine. Yet, had we the proof of but one experiment that any undivided particle, in breaking a hard and solid body, suffered a division, we might by virtue of this rule conclude that the undivided as well as the divided particles may be divided and actually separated to infinity.

Lastly, if it universally appears, by experiments and astronomical observations, that all bodies about the earth gravitate towards the earth, and that in proportion to the quantity of matter which they severally contain; that the moon likewise, according to the quantity of its matter, gravitates towards the earth; that, on the other hand, our sea gravitates towards the moon; and all the planets one towards another; and the comets in like manner towards the sun; we must, in consequence of this rule, universally allow that all bodies what-soever are endowed with a principle of mutual gravitation. For the argument from the appearances concludes with more force for the universal gravitation of all bodies than for their impenetrability; of which, among those in the celestial regions, we have no experiments, nor any manner of observation. Not that I affirm gravity to be essen-tial to bodies: by their *vis insita* I mean nothing but their inertia. This is immutable. Their gravity is diminished as they recede from the earth.

Rule IV

In experimental philosophy we are to look upon propositions in-ferred by general induction from phenomena as accurately or very nearly true, notwithstanding any contrary hypotheses that may be imagined, till such time as other phenomena occur, by which they may either be made more accurate, or liable to exceptions.

This rule we must follow, that the argument of induction may not be evaded by hypotheses.

GENERAL SCHOLION

Hitherto we have explained the phenomena of the heavens and of our sea by the power of gravity, but have not yet assigned the cause of this power. This is certain, that it must proceed from a cause that penetrates to the very centres of the sun and planets, without suffering the least diminution of its force; that operates not according to the quantity of the surfaces of the particles upon which it acts (as mechanical causes used to do), but according to the quantity of the solid matter which they contain, and propagates its virtue on all sides to immense distances, decreasing always as the inverse square of the distances. Gravitation towards the sun is made up out of the gravitations towards the several particles of which the body of the sun is composed; and in receding from the sun decreases accurately as the inverse square of the distances as far as the orbit of Saturn, as evidently appears from the quiescence of the aphelion of the planets; nay, and even to the remotest aphelion of the comets, if those aphelions are also quiescent. But hitherto I have not been able to discover the cause of those properties of gravity from phenomena, and I frame no hypotheses; for whatever is not deduced from the phenomena is to be called an hypothesis; and hypotheses, whether metaphysical or physical, whether of occult qualities or mechanical, have no place in experimental philosophy. In this philosophy particular propositions are inferred from the phenomena, and afterwards rendered general by induction. Thus it was that the impenetrability, the mobility, and the impulsive force of bodies, and the laws of motion and of gravitation, were discovered. And to us it is enough that gravity does really exist, and act according to the laws which we have explained, and abundantly serves to account for all the motions of the celestial bodies, and of our sea.

And now we might add something concerning a certain most subtle spirit which pervades and lies hid in all gross bodies; by the force and action of which spirit the particles of bodies attract one another at near distances, and cohere, if contiguous; and electric bodies operate to greater distances, as well repelling as attracting the neighboring corpuscles; and light is emitted, reflected, refracted, inflected, and heats bodies; and all sensation is excited, and the members of animal bodies move at the command of the will, namely, by the vibrations of this spirit, mutually propagated along the solid filaments of the nerves, from the outward organs of sense to the brain, and from the brain into the muscles. But these are things that cannot be explained in few words, nor are we furnished with that sufficiency of experiments which is required to an accurate determination and demonstration of the laws by which this electric and elastic spirit operates.

OPTICKS: QUERIES

All these things being consider'd, it seems probable to me, that God in the Beginning form'd Matter in solid, massy, hard, impenetrable, movement Particles, of such Sizes and Figures, and with such other Properties, and in such Proportion to Space, as most conduced to the End for which he form'd them; and that these primitive Particles being Solids, are incomparably harder than any porous Bodies compounded of them; even so very hard, as never to wear or break in pieces; no ordinary Power being able to divide what God himself made one in the first Creation. While the Particles continue entire, they may compose Bodies of one and the same Nature and Texture in all Ages: But should they wear away, or break in pieces, the Nature of Things depending on them, would be changed. Water and Earth, composed of old worn Particles and Fragments of Particles, would not be of the same Nature and Texture now, with Water and Earth composed of entire Particles in the Beginning. And therefore, that Nature may be lasting, the Changes of corporeal Things are to be placed only in the various Separations and new Associations and Motions of these permanent Particles; compound Bodies being apt to break, not in the midst of solid Particles, but where those Particles are laid together, and only touch in a few Points.

It seems to me farther, that these Particles have not only a *Vis inertiæ*, accompanied with such passive Laws of Motion as naturally result from that Force, but also that they are moved by certain active Principles, such as is that of Gravity, and that which causes Fermentation, and the Cohesion of Bodies. These Principles I consider, not as occult Qualities, supposed to result from the specifick Forms of Things, but as general Laws of Nature, by which the Things themselves are form'd; their Truth appearing to us by Phænomena, though their Causes be not yet discover'd. For these are manifest Qualities, and their Causes only are occult. And the *Aristotelians* gave the Name of occult Qualities, not to manifest

From Sir Isaac Newton, *OPTICKS*, pp. 400–05, by permission of Dover Publications, Inc.

Qualities, but to such Qualities only as they supposed to lie hid in Bodies, and to be the unknown Causes of manifest Effects: Such as would be the Causes of Gravity, and of magnetick and electrick Attractions, and of Fermentations, if we should suppose that these Forces or Actions arose from Qualities unknown to us, and un-capable of being discovered and made manifest. Such occult Quali-ties put a stop to the Improvement of natural Philosophy, and therefore of late Years have been rejected. To tell us that every Species of Things is endow'd with an occult specifick Quality by which it acts and produces manifest Effects, is to tell us nothing: But to derive two or three general Principles of Motion from Phænomena, and afterwards to tell us how the Properties and Ac-tions of all corporeal Things follow from those manifest Principles, would be a very great step in Philosophy, though the Causes of those Principles were not yet discover'd: And therefore I scruple not to propose the Principles of Motion above-mention'd, they being of very general Extent, and leave their Causes to be found out.

Now by the help of these Principles, all material Things seem to have been composed of the hard and solid Particles above-men-tion'd, variously associated in the first Creation by the Counsel of an intelligent Agent. For it became him who created them to set them in order. And if he did so, it's unphilosophical to seek for any other Origin of the World, or to pretend that it might arise out of a Chaos by the mere Laws of Nature; though being once form'd, it may continue by those Laws for many Ages. For while Comets move in very excentrick Orbs in all manner of Positions, blind Fate could never make all the Planets move one and the same way in Orbs concentrick, some inconsiderable Irregularities excepted, which may have risen from the mutual Actions of Comets and Planets upon one another, and which will be apt to increase, till this System wants a Reformation. Such a wonderful Uniformity in the Planetary System must be allowed the Effect of Choice. And so must the Uniformity in the Bodies of Animals, they having gen-erally a right and a left side shaped alike, and on either side of their Bodies two Legs behind, and either two Arms, or two Legs, or two Wings before upon their Shoulders, and between their Shoulders a Neck running down into a Back-bone, and a Head upon it; and in the Head two Ears, two Eyes, a Nose, a Mouth, and a Tongue, alike situated. Also the first Contrivance of those very artificial Parts of Animals, the Eyes, Ears, Brain, Muscles, Heart, Lungs, Midriff, Glands, Larynx, Hands, Wings, swimming Bladders, natural Spectacles, and other Organs of Sense and Motion; and the Instinct of Brutes and Insects, can be the effect of nothing else than the Wisdom and Skill of a powerful ever-living Agent, who being in all Places, is more able by his Will to move the Bodies within his boundless uniform Sensorium, and thereby to form and reform the Parts of the Universe, than we are by our Will to move the Parts of our own Bodies. And yet we are not to consider the World as the Body of God, or the several Parts thereof, as the Parts of God. He is an uniform Being, void of Organs, Members or Parts and they are his Creatures subordinate to him, and subservient to his Will; and he is no more the Soul of them, than the Soul of Man is the Soul of the Species of Things carried through the Organs of Sense into the place of its Sensation, where it perceives them by means of its immediate Presence, without the Intervention of any third thing. The Organs of Sense are not for enabling the Soul to perceive the Species of Things in its Sensorium, but only for conveying them thither; and God has no need of such Organs, he being everywhere present to the Things themselves. And since Space is divisible *in infinitum*, and Matter is not necessarily in all places, it may be also allow'd that God is able to create Particles of Matter of several Sizes and Figures, and in several Proportions to Space, and perhaps of different Densities and Forces, and thereby to vary the Laws of Nature, and make Worlds of several sorts in several Parts of the Universe. At least, I see nothing of Contradiction in all this.

As in Mathematicks, so in Natural Philosophy, the Investigation of difficult Things by the Method of Analysis, ought ever to precede the Method of Composition. This Analysis consists in making Experiments and Observations, and in drawing general Conclusions from them by Induction, and admitting of no Objections against the Conclusions, but such as are taken from Experiments, or other certain Truths. For Hypotheses are not to be regarded in experimental Philosophy. And although the arguing from Experiments and Observations by Induction be no Demonstration of general Conclusions; yet it is the best way of arguing which the Nature of Things admits of, and may be looked upon as so much the stronger, by how much the Induction is more general. And if no Exception occur from Phænomena, the Conclusion may be pronounced generally. But if at any time afterwards any Exception shall occur from Experiments, it may then begin to be pronounced with such Exceptions as occur. By this way of Analysis we may proceed from Compounds to Ingredients, and from Motions to the Forces producing them; and in general, from Effects to their Causes, and from particular Causes to more general ones, till the Argument end in the most general. This is the Method of Analysis: And the Synthesis consists in assuming the Causes discover'd, and establish'd as Principles, and by them explaining the Phænomena proceeding from them, and proving the Explanations.

Rudolf Carnap

Rudolf Carnap, though a German by birth, came to philosophical prominence through his participation in the Vienna Circle, the group of philosophers and philosopher-scientists surrounding Moritz Schlick at the University of Vienna. Since 1928, the year in which he published two basic works, *Der Logische Aufbau der Welt* and *Scheinprobleme in der Philosophie,* Carnap has been in the forefront of technical developments in the philosophy of science and applied logic. So notable have been his contributions that even those who disagree with his conclusions have usually felt obliged to accept his statement of a problem as their point of departure and to use the technical tools Carnap developed as the means for further advancement. Thus, the distinction between internal and external questions developed in the present article has become almost normative in subsequent discussions of formal systems. This article, originally published in 1950, has been reproduced in various anthologies. It is included here because of the central role it plays in contemporary discussions of the epistemological problems concerned with scientific realism. Although Carnap retired from his teaching position at the University of California in 1961 he remained active in the philosophy of science until his death in 1970.

Empiricism, Semantics, and Ontology

1. The Problem of Abstract Entities

Empiricists are in general rather suspicious with respect to any kind of abstract entities like properties, classes, relations, numbers, propositions, etc. They usually feel much more in sympathy with nominalists than with realists (in the medieval sense). As far as possible they try to avoid any reference to abstract entities and to restrict themselves to what is sometimes called a nominalistic language, i.e., one not containing such references. However, within certain scientific contexts it seems hardly possible to avoid them. In the case of mathematics, some empiricists try to find a way out by treating the whole of mathematics as a mere calculus, a formal system for which no interpretation is given or can be given. Accordingly, the mathematician is said to speak not about numbers, functions, and infinite classes, but merely about meaningless symbols and formulas manipulated according to given formal rules. In physics it is more difficult to shun the suspected entities, because the language of physics serves for the communication of reports and predictions and hence cannot be taken as a mere calculus. A physicist who is suspicious of abstract entities may perhaps try to declare a certain part of the language of physics as uninterpreted and uninterpretable, that part which refers to real numbers as space-time coordinates or as values of physical magnitudes, to functions, limits, etc. More probably he will just speak about all these things like anybody else but with an uneasy conscience, like a man who in his everyday life does with qualms many things which are not in accord with the high moral principles he professes on Sundays. Recently the problem of abstract

From Rudolf Carnap, MEANING AND NECESSITY, pp. 205–21, © 1947 and 1956 by The University of Chicago. All rights reserved.

entities has arisen again in connection with semantics, the theory of meaning and truth. Some semanticists say that certain expressions designate certain entities, and among these designated entities they include not only concrete material things but also abstract entities, e.g., properties as designated by predicates and propositions as designated by sentences.[1] Others object strongly to this procedure as violating the basic principles of empiricism and leading back to a metaphysical ontology of the Platonic kind.

It is the purpose of this article to clarify this controversial issue. The nature and implications of the acceptance of a language referring to abstract entities will first be discussed in general; it will be shown that using such a language does not imply embracing a Platonic ontology but is perfectly compatible with empiricism and strictly scientific thinking. Then the special question of the role of abstract entities in semantics will be discussed. It is hoped that the clarification of the issue will be useful to those who would like to accept abstract entities in their work in mathematics, physics, semantics, or any other field; it may help them to overcome nominalistic scruples.

2. Linguistic Frameworks

Are there properties, classes, numbers, propositions? In order to understand more clearly the nature of these and related problems, it is above all necessary to recognize a fundamental distinction between two kinds of questions concerning the existence or reality of entities. If someone wishes to speak in his language about a new kind of entities, he has to introduce a system of new ways of speaking, subject to new rules; we shall call this procedure the construction of a linguistic *framework* for the new entities in question. And now we must distinguish two kinds of questions of existence: first, questions of the existence of certain entities of the new kind *within the framework;* we call them *internal questions;* and second, questions concerning the existence or reality *of the system of entities as a whole,* called *external questions.* Internal questions and possible answers to them are formulated with the help of the new forms of

1. The terms 'sentence' and 'statement' are here used synonymously for declarative (indicative, propositional) sentences.

expressions. The answers may be found either by purely logical methods or by empirical methods, depending upon whether the framework is a logical or a factual one. An external question is of a problematic character which is in need of closer examination.

The World of Things

Let us consider as an example the simplest kind of entities dealt with in the everyday language: the spatio-temporally ordered system of observable things and events. Once we have accepted the thing language with its framework for things, we can raise and answer internal questions, e.g., "Is there a white piece of paper on my desk?," "Did King Arthur actually live?," "Are unicorns and centaurs real or merely imaginary?," and the like. These questions are to be answered by empirical investigations. Results of observations are evaluated according to certain rules as confirming or disconfirming evidence for possible answers. (This evaluation is usually carried out, of course, as a matter of habit rather than a deliberate, rational procedure. But it is possible, in a rational reconstruction, to lay down explicit rules for the evaluation. This is one of the main tasks of a pure, as distinguished from a psychological, epistemology.) The concept of reality occurring in these internal questions is an empirical, scientific, non-metaphysical concept. To recognize something as a real thing or event means to succeed in incorporating it into the system of things at a particular space-time position so that it fits together with the other things recognized as real, according to the rules of the framework.

From these questions we must distinguish the external question of the reality of the thing world itself. In contrast to the former questions, this question is raised neither by the man in the street nor by scientists, but only by philosophers. Realists give an affirmative answer, subjective idealists a negative one, and the controversy goes on for centuries without ever being solved. And it cannot be solved because it is framed in a wrong way. To be real in the scientific sense means to be an element of the system; hence this concept cannot be meaningfully applied to the system itself. Those who raise the question of the reality of the thing world itself have perhaps in mind not a theoretical question as their formulation seems to suggest, but rather a practical question, a matter of a practical decision concerning the structure of our language. We have to make the choice whether or not to accept and use the forms of expression in the framework in question.

In the case of this particular example, there is usually no deliberate choice because we all have accepted the thing language early in our lives as a matter of course. Nevertheless, we may regard it as a matter of decision in this sense: we are free to choose to continue using the thing language or not; in the latter case we could restrict ourselves to a language of sense-data and other "phenomenal" entities, or construct an alternative to the customary thing language with another structure, or, finally, we could refrain from speaking. If someone decides to accept the thing language, there is no objection against saying that he has accepted the world of things. But this must not be interpreted as if it meant his acceptance of a *belief* in the reality of the thing world; there is no such belief or assertion or assumption, because it is not a theoretical question. To accept the thing world means nothing more than to accept a certain form of language, in other words, to accept rules for forming statements and for testing, accepting, or rejecting them. The acceptance of the thing language leads, on the basis of observations made, also to the acceptance, belief, and assertion of certain statements. But the thesis of the reality of the thing world cannot be among these statements, because it cannot be formulated in the thing language or, it seems, in any other theoretical language.

The decision of accepting the thing language, although itself not of a cognitive nature, will nevertheless usually be influenced by theoretical knowledge, just like any other deliberate decision concerning the acceptance of linguistic or other rules. The purposes for which the language is intended to be used, for instance, the purpose of communicating factual knowledge, will determine which factors are relevant for the decision. The efficiency, fruitfulness, and simplicity of the use of the thing language may be among the decisive factors. And the questions concerning these qualities are indeed of a theoretical nature. But these questions cannot be identified with the question of realism. They are not yes-no questions but questions of degree. The thing language in the customary form works indeed with a high degree of efficiency for most purposes of everyday life. This is a matter of fact, based upon the content of our experiences. However, it would be wrong to describe this situation by saying:

"The fact of the efficiency of the thing language is confirming evidence for the reality of the thing world"; we should rather say instead: "This fact makes it advisable to accept the thing language."

The System of Numbers

As an example of a system which is of a logical rather than a factual nature let us take the system of natural numbers. The framework for this system is constructed by introducing into the language new expressions with suitable rules: (1) numerals like "five" and sentence forms like "there are five books on the table"; (2) the general term "number" for the new entities, and sentence forms like "five is a number"; (3) expressions for properties of numbers (e.g., "odd," "prime"), relations (e.g., "greater than"), and functions (e.g., "plus"), and sentence forms like "two plus three is five"; (4) numerical variables ("m," "n," etc.) and quantifiers for universal sentences ("for every n, . . .") and existential sentences ("there is an n such that . . .") with the customary deductive rules.

Here again there are internal questions, e.g., "Is there a prime number greater than a hundred?" Here, however, the answers are found, not by empirical investigation based on observations, but by logical analysis based on the rules for the new expressions. Therefore the answers are here analytic, i.e., logically true.

What is now the nature of the philosophical question concerning the existence or reality of numbers? To begin with, there is the internal question which, together with the affirmative answer, can be formulated in the new terms, say, by "There are numbers" or, more explicitly, "There is an n such that n is a number." This statement follows from the analytic statement "five is a number" and is therefore itself analytic. Moreover, it is rather trivial (in contradistinction to a statement like "There is a prime number greater than a million," which is likewise analytic but far from trivial), because it does not say more than that the new system is not empty; but this is immediately seen from the rule which states that words like "five" are substitutable for the new variables. Therefore nobody who meant the question "Are there numbers?" in the internal sense would either assert or even seriously consider a negative answer. This makes it plausible to assume that those philosophers who treat the question of the existence of numbers as a serious philosophical problem and offer lengthy arguments on either side, do not have in mind the internal question. And, indeed, if we were to ask them: "Do you mean the question as to whether the framework of numbers, *if* we were to accept it, would be found to be empty or not?," they would probably reply: "Not at all; we mean a question *prior* to the acceptance of the new framework." They might try to explain what they mean by saying that it is a question of the ontological status of numbers; the question whether or not numbers have a certain metaphysical characteristic called reality (but a kind of ideal reality, different from the material reality of the thing world) or subsistence or status of "independent entities." Unfortunately, these philosophers have so far not given a formulation of their question in terms of the common scientific language. Therefore our judgment must be that they have not succeeded in giving to the external question and to the possible answers any cognitive content. Unless and until they supply a clear cognitive interpretation, we are justified in our suspicion that their question is a pseudo-question, that is, one disguised in the form of a theoretical question while in fact it is nontheoretical; in the present case it is the practical problem whether or not to incorporate into the language the new linguistic forms which constitute the framework of numbers.

The System of Propositions

New variables, "p," "q," etc., are introduced with a rule to the effect that any (declarative) sentence may be substituted for a variable of this kind; this includes, in addition to the sentences of the original thing language, also all general sentences with variables of any kind which may have been introduced into the language. Further, the general term "proposition" is introduced. "p is a proposition" may be defined by "p or not p" (or by any other sentence form yielding only analytic sentences). Therefore, every sentence of the form ". . . is a proposition" (where any sentence may stand in the place of the dots) is analytic. This holds, for example, for the sentence:

(*a*)   "Chicago is large is a proposition."

(We disregard here the fact that the rules of English grammar require not a sentence but a that-clause as the subject of another sentence; accordingly, instead of (*a*) we should have to say "That

Chicago is large is a proposition.") Predicates may be admitted whose argument expressions are sentences; these predicates may be either extensional (e.g., the customary truth-functional connectives) or not (e.g., modal predicates like "possible," "necessary," etc.). With the help of the new variables, general sentences may be formed, e.g.,

- (*b*) "For every *p*, either *p* or not-*p*."
- (*c*) "There is a *p* such that *p* is not necessary and not-*p* is not necessary."
- (*d*) "There is a *p* such that *p* is a proposition."

(*c*) and (*d*) are internal assertions of existence. The statement "There are propositions" may be meant in the sense of (*d*); in this case it is analytic (since it follows from (*a*)) and even trivial. If, however, the statement is meant in an external sense, then it is non-cognitive.

It is important to notice that the system of rules for the linguistic expressions of the propositional framework (of which only a few rules have here been briefly indicated) is sufficient for the introduction of the framework. Any further explanations as to the nature of the propositions (i.e., the elements of the system indicated, the values of the variables "*p*," "*q*," etc.) are theoretically unnecessary because, if correct, they follow from the rules. For example, are propositions mental events (as in Russell's theory)? A look at the rules shows us that they are not, because otherwise existential statements would be of the form: "If the mental state of the person in question fulfils such and such conditions, then there is a *p* such that . . . ." The fact that no references to mental conditions occur in existential statements (like (*c*), (*d*), etc.) shows that propositions are not mental entities. Further, a statement of the existence of linguistic entities (e.g., expressions, classes of expressions, etc.) must contain a reference to a language. The fact that no such reference occurs in the existential statements here, shows that propositions are not linguistic entities. The fact that in these statements no reference to a subject (an observer or knower) occurs (nothing like: "There is a *p* which is necessary for Mr. *X*"), shows that the propositions (and their properties, like necessity, etc.) are not subjective. Although characterizations of these or similar kinds are, strictly speaking, unnecessary, they may nevertheless be practically useful. If they are given, they should be understood, not as ingredient parts of the system, but merely as marginal notes with the purpose of supplying to the reader helpful hints or convenient pictorial associations which may make his learning of the use of the expressions easier than the bare system of the rules would do. Such a characterization is analogous to an extra-systematic explanation which a physicist sometimes gives to the beginner. He might, for example, tell him to imagine the atoms of a gas as small balls rushing around with great speed, or the electromagnetic field and its oscillations as quasi-elastic tensions and vibrations in an ether. In fact, however, all that can accurately be said about atoms or the field is implicitly contained in the physical laws of the theories in question.[2]

The System of Thing Properties

The thing language contains words like "red," "hard," "stone," "house," etc., which are used for describing what things are like. Now we may introduce new variables, say "f," "g," etc., for which those words are substitutable and furthermore the general term

2. In my book *Meaning and Necessity* (Chicago, 1947) I have developed a semantical method which takes propositions as entities designated by sentences (more specifically, as intensions of sentences). In order to facilitate the understanding of the systematic development, I added some informal, extra-systematic explanations concerning the nature of propositions. I said that the term "proposition" "is used neither for a linguistic expression nor for a subjective, mental occurrence, but rather for something objective that may or may not be exemplified in nature. . . . We apply the term 'proposition' to any entities of a certain logical type, namely, those that may be expressed by (declarative) sentences in a language" (p. 27). After some more detailed discussions concerning the relation between propositions and facts, and the nature of false propositions, I added: "It has been the purpose of the preceding remarks to facilitate the understanding of our conception of propositions. If, however, a reader should find these explanations more puzzling than clarifying, or even unacceptable, he may disregard them" (p. 31) (that is, disregard these extra-systematic explanations, not the whole theory of the propositions as intensions of sentences, as one reviewer understood). In spite of this warning, it seems that some of those readers who were puzzled by the explanations, did not disregard them but thought that by raising objections against them they could refute the theory. This is analogous to the procedure of some laymen who by (correctly) criticizing the ether picture or other visualizations of physical theories, thought they had refuted those theories. Perhaps the discussions in the present paper will help in clarifying the role of the system of linguistic rules for the introduction of a framework for entities on the one hand, and that of extra-systematic explanations concerning the nature of the entities on the other.

"property." New rules are laid down which admit sentences like "Red is a property," "Red is a color," "These two pieces of paper have at least one color in common" (i.e., "There is an f such that f is a color, and . . ."). The last sentence is an internal assertion. It is of an empirical, factual nature. However, the external statement, the philosophical statement of the reality of properties—a special case of the thesis of the reality of universals—is devoid of cognitive content.

The Systems of Integers and Rational Numbers

Into a language containing the framework of natural numbers we may introduce first the (positive and negative) integers as relations among natural numbers and then the rational numbers as relations among integers. This involves introducing new types of variables, expressions substitutable for them, and the general terms "integer" and "rational number."

The System of Real Numbers

On the basis of the rational numbers, the real numbers may be introduced as classes of a special kind (segments) of rational numbers (according to the method developed by Dedekind and Frege). Here again a new type of variables is introduced, expressions substitutable for them (e.g., "$\sqrt{2}$"), and the general term "real number."

The Spatio-Temporal Coordinate System for Physics

The new entities are the space-time points. Each is an ordered quadruple of four real numbers, called its coordinates, consisting of three spatial and one temporal coordinates. The physical state of a spatio-temporal point or region is described either with the help of qualitative predicates (e.g., "hot") or by ascribing numbers as values of a physical magnitude (e.g., mass, temperature, and the like). The step from the system of things (which does not contain space-time points but only extended objects with spatial and temporal relations between them) to the physical coordinate system is again a matter of decision. Our choice of certain features, although itself not theoretical, is suggested by theoretical knowledge, either logical or factual. For example, the choice of real numbers rather than rational numbers or integers as coordinates is not much influenced by the facts of experience but mainly due to considerations of mathematical simplicity. The restriction to rational coordinates would not be in conflict with any experimental knowledge we have, because the result of any measurement is a rational number. However, it would prevent the use of ordinary geometry (which says, e.g., that the diagonal of a square with the side 1 has the irrational value $\sqrt{2}$) and thus lead to great complications. On the other hand, the decision to use three rather than two or four spatial coordinates is strongly suggested, but still not forced upon us, by the result of common observations. If certain events allegedly observed in spiritualistic séances, e.g., a ball moving out of a sealed box, were confirmed beyond any reasonable doubt, it might seem advisable to use four spatial coordinates. Internal questions are here, in general, empirical questions to be answered by empirical investigations. On the other hand, the external questions of the reality of physical space and physical time are pseudo-questions. A question like "Are there (really) space-time points?" is ambiguous. It may be meant as an internal question; then the affirmative answer is, of course, analytic and trivial. Or it may be meant in the external sense: "Shall we introduce such and such forms into our language?"; in this case it is not a theoretical but a practical question, a matter of decision rather than assertion, and hence the proposed formulation would be misleading. Or finally, it may be meant in the following sense: "Are our experiences such that the use of the linguistic forms in question will be expedient and fruitful?" This is a theoretical question of a factual, empirical nature. But it concerns a matter of degree; therefore a formulation in the form "real or not?" would be inadequate.

3. What Does Acceptance of a Kind of Entities Mean?

Let us now summarize the essential characteristics of situations involving the introduction of a new kind of entities, characteristics which are common to the various examples outlined above.

The acceptance of a new kind of entities is represented in the language by the introduction of a framework of new forms of expressions to be used according to a new set of rules. There may be new names for particular entities of the kind in question; but some such names may already occur in the language before the introduction of the new framework. (Thus, for example, the thing language contains certainly words of the type of "blue" and "house" before the framework of properties is introduced; and it may contain words like "ten" in sentences of the form "I have ten fingers" before the framework of numbers is introduced.) The latter fact shows that the occurrence of constants of the type in question—regarded as names of entities of the new kind after the new framework is introduced—is not a sure sign of the acceptance of the new kind of entities. Therefore the introduction of such constants is not to be regarded as an essential step in the introduction of the framework. The two essential steps are rather the following. First, the introduction of a general term, a predicate of higher level, for the new kind of entities, permitting us to say of any particular entity that it belongs to this kind (e.g., "Red is a *property*," "Five is a *number*"). Second, the introduction of variables of the new type. The new entities are values of these variables; the constants (and the closed compound expressions, if any) are substitutable for the variables.[3] With the help of the variables, general sentences concerning the new entities can be formulated.

After the new forms are introduced into the language, it is possible to formulate with their help internal questions and possible answers to them. A question of this kind may be either empirical or logical; accordingly a true answer is either factually true or analytic.

From the internal questions we must clearly distinguish external questions, i.e., philosophical questions concerning the existence or reality of the total system of the new entities. Many philosophers regard a question of this kind as an ontological question which must be raised and answered *before* the introduction of the new language forms. The latter introduction, they believe, is legitimate only if it can be justified by an ontological insight supplying an affirmative answer to the question of reality. In contrast to this view, we take the position that the introduction of the new ways of speaking does not need any theoretical justification because it does not imply any

3. W. V. Quine was the first to recognize the importance of the introduction of variables as indicating the acceptance of entities. "The ontology to which one's use of language commits him comprises simply the objects that he treats as falling . . . within the range of values of his variables" ([Notes], p. 118; compare also his [Designation] and [Universals]).

assertion of reality. We may still speak (and have done so) of "the acceptance of the new entities" since this form of speech is customary; but one must keep in mind that this phrase does not mean for us anything more than acceptance of the new framework, i.e., of the new linguistic forms. Above all, it must not be interpreted as referring to an assumption, belief, or assertion of "the reality of the entities." There is no such assertion. An alleged statement of the reality of the system of entities is a pseudo-statement without cognitive content. To be sure, we have to face at this point an important question; but it is a practical, not a theoretical question; it is the question of whether or not to accept the new linguistic forms. The acceptance cannot be judged as being either true or false because it is not an assertion. It can only be judged as being more or less expedient, fruitful, conducive to the aim for which the language is intended. Judgments of this kind supply the motivation for the decision of accepting or rejecting the kind of entities.[4]

Thus it is clear that the acceptance of a linguistic framework must not be regarded as implying a metaphysical doctrine concerning the reality of the entities in question. It seems to me due to a neglect of this important distinction that some contemporary nominalists label the admission of variables of abstract types as "Platonism."[5] This is, to say the least, an extremely misleading terminology. It leads to the absurd consequence, that the position of everybody who accepts the language of physics with its real number variables

4. For a closely related point of view on these questions see the detailed discussions in Herbert Feigl, "Existential Hypotheses," *Philosophy of Science,* 17 (1950), 35–62.

5. Paul Bernays, "Sur le platonisme dans les mathématiques" (*L'Enseignement math.,* 34 (1935), 52–69). W. V. Quine, see previous footnote and a recent paper [What]. Quine does not acknowledge the distinction which I emphasize above, because according to his general conception there are no sharp boundary lines between logical and factual truth, between questions of meaning and questions of fact, between the acceptance of a language structure and the acceptance of an assertion formulated in the language. This conception, which seems to deviate considerably from customary ways of thinking, will be explained in his article [Semantics]. When Quine in the article [What] classifies my logicistic conception of mathematics (derived from Frege and Russell) as "platonic realism" (p. 33), this is meant (according to a personal communication from him) not as ascribing to me agreement with Plato's metaphysical doctrine of universals, but merely as referring to the fact that I accept a language of mathematics containing variables of higher levels. With respect to the basic attitude to take in choosing a language form (an 'ontology' in Quine's terminology, which seems to me misleading), there appears now to be agreement between us: "the obvious counsel is tolerance and an experimental spirit" ([What], p. 38).

(as a language of communication, not merely as a calculus) would be called Platonistic, even if he is a strict empiricist who rejects Platonic metaphysics.

A brief historical remark may here be inserted. The non-cognitive character of the questions which we have called here external questions was recognized and emphasized already by the Vienna Circle under the leadership of Moritz Schlick, the group from which the movement of logical empiricism originated. Influenced by ideas of Ludwig Wittgenstein, the Circle rejected both the thesis of the reality of the external world and the thesis of its irreality as pseudo-statements; [6] the same was the case for both the thesis of the reality of universals (abstract entities, in our present terminology) and the nominalistic thesis that they are not real and that their alleged names are not names of anything but merely *flatus voci*. (It is obvious that the apparent negation of a pseudo-statement must also be a pseudo-statement.) It is therefore not correct to classify the members of the Vienna Circle as nominalists, as is sometimes done. However, if we look at the basic anti-metaphysical and pro-scientific attitude of most nominalists (and the same holds for many materialists and realists in the modern sense), disregarding their occasional pseudo-theoretical formulations, then it is, of course, true to say that the Vienna Circle was much closer to those philosophers than to their opponents.

4. Abstract Entities in Semantics

The problem of the legitimacy and the status of abstract entities has recently again led to controversial discussions in connection with semantics. In a semantical meaning analysis certain expressions in a language are often said to designate (or name or denote or signify or refer to) certain extra-linguistic entities.[7] As long as physical

6. See Carnap, *Scheinprobleme in der Philosophie; das Fremd psychische und der Realismusstreit,* Berlin, 1928. Moritz Schlick, *Positivismus und Realismus,* reprinted in *Gesammelte Aufsätze,* Wien, 1938.

7. See [I]; *Meaning and Necessity* (Chicago, 1947). The distinction I have drawn in the latter book between the method of the name-relation and the method of intension and extension is not essential for our present discussion. The term 'designation' is used in the present article in a neutral way; it may be understood as referring to the name-relation or to the intension-relation or to the extension-relation or to any similar relations used in other semantical methods.

things or events (e.g., Chicago or Caesar's death) are taken as designata (entities designated), no serious doubts arise. But strong objections have been raised, especially by some empiricists, against abstract entities as designata, e.g., against semantical statements of the following kind:

(1) "The word 'red' designates a property of things";
(2) "The word 'color' designates a property of properties of things";
(3) "The word 'five' designates a number";
(4) "The word 'odd' designates a property of numbers";
(5) "The sentence 'Chicago is large' designates a proposition."

Those who criticize these statements do not, of course, reject the use of the expressions in question, like "red" or "five"; nor would they deny that these expressions are meaningful. But to be meaningful, they would say, is not the same as having a meaning in the sense of an entity designated. They reject the belief, which they regard as implicitly presupposed by those semantical statements, that to each expression of the types in question (adjectives like "red," numerals like "five," etc.) there is a particular real entity to which the expression stands in the relation of designation. This belief is rejected as incompatible with the basic principles of empiricism or of scientific thinking. Derogatory labels like "Platonic realism," "hypostatization," or "'Fido'-Fido principle" are attached to it. The latter is the name given by Gilbert Ryle [Meaning] to the criticized belief, which, in his view, arises by a naïve inference of analogy: just as there is an entity well known to me, viz. my dog Fido, which is designated by the name "Fido," thus there must be for every meaningful expression a particular entity to which it stands in the relation of designation or naming, i.e., the relation exemplified by "Fido"-Fido. The belief criticized is thus a case of hypostatization, i.e., of treating as names expressions which are not names. While "Fido" is a name, expressions like "red," "five," etc., are said not to be names, not to designate anything.

Our previous discussion concerning the acceptance of frameworks enables us now to clarify the situation with respect to abstract entities as designata. Let us take as an example the statement:

(a) "'Five' designates a number."

The formulation of this statement presupposes that our language L contains the forms of expressions which we have called the framework of numbers, in particular, numerical variables and the general term "number." If L contains these forms, the following is an analytic statement in L:

   (b)   "Five is a number."

Further, to make the statement (a) possible, L must contain an expression like "designates" or "is a name of" for the semantical relation of designation. If suitable rules for this term are laid down, the following is likewise analytic:

   (c)   " 'Five' designates five."

(Generally speaking, any expression of the form " '. . .' designates . . ." is an analytic statement provided the term ". . ." is a constant in an accepted framework. If the latter condition is not fulfilled, the expression is not a statement.) Since (a) follows from (c) and (b), (a) is likewise analytic.

Thus it is clear that *if* someone accepts the framework of numbers, then he must acknowledge (c) and (b) and hence (a) as true statements. Generally speaking, if someone accepts a framework for a certain kind of entities, then he is bound to admit the entities as possible designata. Thus the question of the admissibility of entities of a certain type or of abstract entities in general as designata is reduced to the question of the acceptability of the linguistic framework for those entities. Both the nominalistic critics, who refuse the status of designators or names to expressions like "red," "five," etc., because they deny the existence of abstract entities, and the skeptics, who express doubts concerning the existence and demand evidence for it, treat the question of existence as a theoretical question. They do, of course, not mean the internal question; the affirmative answer to *this* question is analytic and trivial and too obvious for doubt or denial, as we have seen. Their doubts refer rather to the system of entities itself; hence they mean the external question. They believe that only after making sure that there really is a system of entities of the kind in question are we justified in accepting the framework by incorporating the linguistic forms into our language. However, we have seen that the external question is not a theoretical question but rather the practical question whether or not to accept those linguistic forms. This acceptance is not in need of a theoretical justification (except with respect to expediency and fruit-fulness), because it does not imply a belief or assertion. Ryle says that the "Fido"-Fido principle is "a grotesque theory." Grotesque or not, Ryle is wrong in calling it a theory. It is rather the practical decision to accept certain frameworks. Maybe Ryle is historically right with respect to those whom he mentions as previous repre-sentatives of the principle, viz. John Stuart Mill, Frege, and Russell. If these philosophers regarded the acceptance of a system of entities as a theory, an assertion, they were victims of the same old, meta-physical confusion. But it is certainly wrong to regard *my* semantical method as involving a belief in the reality of abstract entities, since I reject a thesis of this kind as a metaphysical pseudo-statement.

The critics of the use of abstract entities in semantics overlook the fundamental difference between the acceptance of a system of en-tities and an internal assertion, e.g., an assertion that there are elephants or electrons or prime numbers greater than a million. Whoever makes an internal assertion is certainly obliged to justify it by providing evidence, empirical evidence in the case of electrons, logical proof in the case of the prime numbers. The demand for a theoretical justification, correct in the case of internal assertions, is sometimes wrongly applied to the acceptance of a system of entities. Thus, for example, Ernest Nagel in [Review C.] asks for "evidence relevant for affirming with warrant that there are such entities as infinitesimals or propositions." He characterizes the evidence re-quired in these cases—in distinction to the empirical evidence in the case of electrons—as "in the broad sense logical and dialectical." Beyond this no hint is given as to what might be regarded as rele-vant evidence. Some nominalists regard the acceptance of abstract entities as a kind of superstition or myth, populating the world with fictitious or at least dubious entities, analogous to the belief in cen-taurs or demons. This shows again the confusion mentioned, because a superstition or myth is a false (or dubious) internal statement.

Let us take as example the natural numbers as cardinal numbers, i.e., in contexts like "Here are three books." The linguistic forms of the framework of numbers, including variables and the general term "number," are generally used in our common language of communication; and it is easy to formulate explicit rules for their use. Thus the logical characteristics of this framework are suffi-ciently clear (while many internal questions, i.e., arithmetical questions, are, of course, still open). In spite of this, the controversy concerning the external question of the ontological reality of the system of numbers continues. Suppose that one philosopher says: "I believe that there are numbers as real entities. This gives me the right to use the linguistic forms of the numerical framework and to make semantical statements about numbers as designata of numerals." His nominalistic opponent replies: "You are wrong; there are no numbers. The numerals may still be used as meaningful expressions. But they are not names, there are no entities designated by them. Therefore the word 'number' and numerical variables must not be used (unless a way were found to introduce them as merely abbreviating devices, a way of translating them into the nominalistic thing language)." I cannot think of any possible evidence that would be regarded as relevant by both philosophers, and therefore, if actually found, would decide the controversy or at least make one of the opposite theses more probable than the other. (To construe the numbers as classes or properties of the second level, according to the Frege-Russell method, does, of course, not solve the controversy, because the first philosopher would affirm and the second deny the existence of the system of classes or properties of the second level.) Therefore I feel compelled to regard the external question as a pseudo-question, until both parties to the controversy offer a common interpretation of the question as a cognitive question; this would involve an indication of possible evidence regarded as relevant by both sides.

There is a particular kind of misinterpretation of the acceptance of abstract entities in various fields of science in semantics, that needs to be cleared up. Certain early British empiricists (e.g., Berkeley and Hume) denied the existence of abstract entities on the ground that immediate experience presents us only with particulars, not with universals, e.g., with this red patch, but not with Redness or Color-in-General; with this scalene triangle, but not with Scalene Triangularity or Triangularity-in-General. Only entities belonging to a type of which examples were to be found within immediate experience could be accepted as ultimate constituents of reality. Thus, according to this way of thinking, the existence of abstract entities could be asserted only if one could show either that some abstract entities fall within the given, or that abstract entities can be defined in terms of the types of entity which are given. Since these empiricists found no abstract entities within the realm of sense-data, they either denied their existence, or else made a futile attempt to define universals in terms of particulars. Some contemporary philosophers, especially English philosophers following Bertrand Russell, think in basically similar terms. They emphasize a distinction between the data (that which is immediately given in consciousness, e.g., sense-data, immediately past experiences, etc.) and the constructs based on the data. Existence or reality is ascribed only to the data; the constructs are not real entities; the corresponding linguistic expressions are merely ways of speech not actually designating anything (reminiscent of the nominalists' *flatus vocis*). We shall not criticize here this general conception. (As far as it is a principle of accepting certain entities and not accepting others, leaving aside any ontological, phenomenalistic and nominalistic pseudo-statements, there cannot be any theoretical objection to it.) But if this conception leads to the view that other philosophers or scientists who accept abstract entities thereby assert or imply their occurrence as immediate data, then such a view must be rejected as a misinterpretation. References to space-time points, the electromagnetic field, or electrons in physics, to real or complex numbers and their functions in mathematics, to the excitatory potential or unconscious complexes in psychology, to an inflationary trend in economics, and the like, do not imply the assertion that entities of these kinds occur as immediate data. And the same holds for references to abstract entities as designata in semantics. Some of the criticisms by English philosophers against such references give the impression that, probably due to the misinterpretation just indicated, they accuse the semanticist not so much of bad metaphysics (as some nominalists would do) but of bad psychology. The fact that they regard a semantical method involving abstract entities not merely as doubtful and perhaps wrong, but as manifestly absurd, preposterous and grotesque, and that they show a deep horror and indignation against this method, is perhaps to be explained by a misinterpretation of the kind described. In fact, of course, the semanticist does not in the least assert or imply that the abstract entities to which he refers can be experienced as immediately given either by sensation or by a kind of rational intuition. An assertion of this kind would indeed be very dubious psychology. The psychological question as to which kinds of entities do and which do not occur as immediate data is entirely irrelevant for semantics, just as it is for physics, mathematics, economics, etc., with respect to the examples mentioned above.[8]

5. Conclusion

For those who want to develop or use semantical methods, the decisive question is not the alleged ontological question of the existence of abstract entities but rather the question whether the use of abstract linguistic forms or, in technical terms, the use of variables beyond those for things (or phenomenal data), is expedient and fruitful for the purposes for which semantical analyses are made, viz. the analysis, interpretation, clarification, or construction of languages of communication, especially languages of science. This question is here neither decided nor even discussed. It is not a question simply of yes or no, but a matter of degree. Among those philosophers who have carried out semantical analyses and thought about suitable tools for this work, beginning with Plato and Aristotle and, in a more technical way on the basis of modern logic, with C. S. Peirce and Frege, a great majority accepted abstract entities. This does, of course, not prove the case. After all, semantics in the technical sense is still in the initial phases of its development, and we must be prepared for possible fundamental changes in methods. Let us therefore admit that the nominalistic critics may possibly be right. But if so, they will have to offer better arguments than they have so far. Appeal to ontological insight will not carry much weight. The critics will have to show that it is possible to construct a semantical method which avoids all references to abstract entities and achieves by simpler means essentially the same results as the other methods.

The acceptance or rejection of abstract linguistic forms, just as the acceptance or rejection of any other linguistic forms in any branch of science, will finally be decided by their efficiency as

8. Wilfrid Sellars ("Acquaintance and Description Again," in *Journal of Philos.*, 46 (1949), 496–504; see pp. 502 f.) analyzes clearly the roots of the mistake "of taking the designation relation of semantic theory to be a reconstruction of *being present to an experience*."

instruments, the ratio of the results achieved to the amount and complexity of the efforts required. To decree dogmatic prohibitions of certain linguistic forms instead of testing them by their success or failure in practical use, is worse than futile; it is positively harmful because it may obstruct scientific progress. The history of science shows examples of such prohibitions based on prejudices deriving from religious, mythological, metaphysical, or other irrational sources, which slowed up the developments for shorter or longer periods of time. Let us learn from the lessons of history. Let us grant to those who work in any special field of investigation the freedom to use any form of expression which seems useful to them; the work in the field will sooner or later lead to the elimination of those forms which have no useful function. *Let us be cautious in making assertions and critical in examining them, but tolerant in permitting linguistic forms.*

Carl G. Hempel

Carl G. Hempel, originally trained as a physicist, came to philosophy through the Berlin group, a group quite similar to the more widely known Vienna Circle with which it shared close ties. Since coming to the United States in 1937, Hempel has taught philosophy at Yale and Princeton. His work in the philosophy of science has been characterized by a probing analysis of the inner logic of accepted positions. The clarity and coherence of these analyses have guaranteed his articles a central role in discussions of such problems as: the covering-law or hypothetical-deductive model of scientific explanation, meaning and verification, the paradoxes of confirmation, and the logic of scientific explanation. An article he wrote in conjunction with P. Oppenheim on the logic of scientific explanation has probably influenced philosophers of science more than any other contemporary article in the field. In the present excerpt Hempel summarizes and clarifies ideas developed in the more detailed and technical paper he wrote with Oppenheim.

On the Logic of Explanation

Empirical science, in all its major branches, seeks not only to *describe* the phenomena in the world of our experience, but also to *explain* or *understand* them. While this is widely recognized, it is often held, however, that there exist fundamental differences between the explanatory *methods* appropriate to the different fields of empirical science. In the physical sciences, according to this view, all explanation is achieved ultimately by reference to causal or correlational antecedents; whereas in psychology and the social and historical disciplines—and, according to some, even in biology —the establishment of causal or correlational connections, while desirable and important, is not sufficient. Proper understanding of the phenomena studied in these fields is held to require other types of explanation.

One of the explanatory methods that have been developed for this purpose is that of functional analysis, which has found extensive use in biology, psychology, sociology, and anthropology. This procedure raises problems of considerable interest for the comparative methodology of empirical science. The present essay is an attempt to clarify some of these problems; its object is to examine the logical structure of functional analysis and its explanatory and predictive significance by means of a confrontation with the principal characteristics of the explanatory procedures used in the physical sciences. We begin therefore with a brief examination of the latter.

From pp. 271–277, "The Logic of Functional Analysis," by Carl G. Hempel in SYMPOSIUM ON SOCIOLOGICAL THEORY edited by Llewellyn Gross. Copyright © 1959 by Harper & Row, Publishers, Incorporated. By permission of the publishers. The changes included in the reprinting of this article in Carl G. Hempel, *Aspects of Scientific Explanation* (New York: Free Press, 1965) have been incorporated.

Nomological Explanation: Deductive and Inductive

In a beaker filled to the brim with water at room temperature, there floats a chunk of ice which partly extends above the surface. As the ice gradually melts, one might expect the water in the beaker to overflow. Actually the water level remains unchanged. How is this to be explained? The key to an answer is provided by Archimedes' principle, according to which a solid body floating in a liquid displaces a volume of liquid which has the same weight as the body itself. Hence the chunk of ice has the same weight as the volume of water its submerged portion displaces. Since melting does not affect the weights involved, the water into which the ice turns has the same weight as the ice itself, and hence, the same weight as the water initially displaced by the submerged portion of the ice. Having the same weight, it also has the same volume as the displaced water; hence the melting ice yields a volume of water that suffices exactly to fill the space initially occupied by the submerged part of the ice. Therefore, the water level remains unchanged.

This account (which deliberately disregards certain effects of small magnitude) is an example of an argument intended to explain a given event. Like any explanatory argument, it falls into two parts, which will be called the *explanans* and the *explanandum*.[1] The latter is the statement, or set of statements, describing the phe-

---

1. These terms are given preference over the more familiar words 'explicans' and 'explicandum,' in order to reserve the latter for use in the context of philosophical explication in the technical sense proposed by R. Carnap; see, for example, his *Logical Foundations of Probability* (Chicago: University of Chicago Press, 1950), secs. 1–3. The terms 'explanans' and 'explanandum' were introduced, for this reason, in an earlier article: Carl G. Hempel and P. Oppenheim, "Studies in the Logic of Explanation," *Philosophy of Science*, 15 (1948), pp. 135–75 (reprinted in the present volume). While that article does not deal explicitly with inductive explanation, its first four sections contain various further considerations on deductive explanation that are relevant to the present study. For a careful critical examination of some points of detail discussed in the earlier article, such as especially the relation between explanation and prediction, see the essay by I. Scheffler, "Explanation, Prediction, and Abstraction," *The British Journal for the Philosophy of Science*, 7 (1957), pp. 293–309, which also contains some interesting comments bearing on functional analysis.

nomenon to be explained; the former is the statement, or set of statements, adduced to provide an explanation. In our illustration, the explanandum states that at the end of the process, the beaker contains only water, with its surface at the same level as at the beginning. To explain this, the explanans adduces, first of all, certain laws of physics; among them, Archimedes' principle; laws to the effect that at temperatures above 0°C. and atmospheric pressure, a body of ice turns into a body of water having the same weight; and the law that, at any fixed temperature and pressure, amounts of water that are equal in weight are also equal in volume.

In addition to these laws, the explanans contains a second group of statements; these describe certain particular circumstances which, in the experiment, precede the outcome to be explained; such as the facts that at the beginning, there is a chunk of ice floating in a beaker filled with water; that the water is at room temperature; and that the beaker is surrounded by air at the same temperature and remains undisturbed until the end of the experiment.

The explanatory import of the whole argument lies in showing that the outcome described in the explanandum was to be expected in view of the antecedent circumstances and the general laws listed in the explanans. More precisely, the explanation may be construed as an argument in which the explanandum is deduced from the explanans. Our example then illustrates what we will call explanation by deductive subsumption under general laws, or briefly, *deductive-nomological explanation*. The general form of such an explanation is given by the following schema:

$$(2.1) \qquad \left. \begin{array}{c} L_1, \ L_2, \ \ldots, \ L_m \\[1em] C_1, \ C_2, \ \ldots, \ C_n \end{array} \right\} \text{Explanans}$$

$$\overline{}$$

$$E \qquad \text{Explanandum}$$

Here, $L_1, L_2, \ldots, L_m$ are general laws and $C_1, C_2, \ldots, C_n$ are statements of particular fact; the horizontal line separating the conclusion E from the premises indicates that the former follows logically from the latter.

In our example, the phenomenon to be explained is a particular event that takes place at a certain place and time. But the method of deductive subsumption under general laws lends itself also to the explanation of what might be called "general facts" or uniformities, such as those expressed by laws of nature. For example, the question why Galileo's law holds for physical bodies falling freely near the earth's surface can be answered by showing that the law refers to a special case of accelerated motions under gravitational attraction, and that it can be deduced from the general laws for such motion (namely, Newton's laws of motion and of gravitation) by applying these to the special case where two bodies are involved, one of them the earth and the other the falling object, and where the distance between their centers of gravity equals the length of the earth's radius. Thus, an explanation of the regularities expressed by Galileo's law can be achieved by deducing the latter from the Newtonian laws and from statements specifying the mass and the radius of the earth; the latter two yield the value of the constant acceleration of free fall near the earth.

It might be helpful to mention one further illustration of the role of deductive-nomological explanation in accounting for particular facts as well as for general uniformities or laws. The occurrence of a rainbow on a given occasion can be deductively explained by reference to (1) certain particular determining conditions, such as the presence of raindrops in the air, sunlight falling on these drops, the observer facing away from the sun, etc., and (2) certain general laws, especially those of optical reflection, refraction, and dispersion. The fact that these laws hold can be explained in turn by deduction from the more comprehensive principles of, say, the electromagnetic theory of light.

Thus, the method of deductive-nomological explanation accounts for a particular event by subsuming it under general laws in the manner represented by the schema (2.1); and it can similarly serve to explain the fact that a given law holds by showing that the latter is subsumable, in the same fashion, under more comprehensive laws or theoretical principles. In fact, one of the main objectives of a theory (such as, say, the electromagnetic theory of light) is precisely to provide a set of principles—often expressed in terms of "hypothetical," not directly observable, entities (such as electric and magnetic field vectors)—which will deductively account for a group of antecedently established "empirical generalizations" (such as the laws of rectilinear propagation, reflection, and refraction of light). Frequently, a theoretical explanation will show that the empirical generalizations hold only approximately. For example, the application of Newtonian theory to free fall near the earth yields a law that is like Galileo's except that the acceleration of the fall is seen not to be strictly constant, but to vary slightly with geographical location, altitude above sea level, and certain other factors.

The general laws or theoretical principles that serve to account for empirical generalizations may in turn be deductively subsumable under even more comprehensive principles; for example, Newton's theory of gravitation can be subsumed, as an approximation, under that of the general theory of relativity. Obviously, this explanatory hierarchy has to end at some point. Thus, at any time in the development of empirical science, there will be certain facts which, at that time, are not explainable; these include the most comprehensive general laws and theoretical principles then known and, of course, many empirical generalizations and particular facts for which no explanatory principles are available at the time. But this does not imply that certain facts are intrinsically unexplainable and thus must remain unexplained forever: any particular fact as yet unexplainable, and any general principle, however comprehensive, may subsequently be found to be explainable by subsumption under even more inclusive principles.

Causal explanation is a special type of deductive-nomological explanation; for a certain event or set of events can be said to have caused a specified "effect" only if there are general laws connecting the former with the latter in such a way that, given a description of the antecedent events, the occurrence of the effect can be deduced with the help of the laws. For example, the explanation of the lengthening of a given iron bar as having been caused by an increase in its temperature amounts to an argument of the form (2.1) whose explanans includes (a) statements specifying the initial length of the bar and indicating that the bar is made of iron and that its temperature was raised, (b) a law pertaining to the increase in the length of any iron bar with rising temperature.[2]

2. An explanation by means of laws which are causal in the technical sense of theoretical physics also has the form (2.1) of a deductive-nomological ex-

Not every deductive-nomological explanation is a causal explanation, however. For example, the regularities expressed by Newton's laws of motion and of gravitation cannot properly be said to *cause* the free fall of bodies near the earth's surface to satisfy Galileo's laws.

Now we must consider another type of explanation, which again accounts for a given phenomenon by reference to general laws, but in a manner which does not fit the deductive pattern (2.1). When little Henry catches the mumps, this might be explained by pointing out that he contracted the disease from a friend with whom he played for several hours just a day before the latter was confined with a severe case of mumps. The particular antecedent factors here invoked are Henry's exposure and, let us assume, the fact that Henry had not had the mumps before. But to connect these with the event to be explained, we cannot adduce a general law to the effect that under the conditions just mentioned, the exposed person invariably contracts the mumps: what can be asserted is only that the disease will be transmitted with high statistical probability. Again, when a neurotic trait in an adult is psychoanalytically explained by reference to critical childhood experiences, the argument explicitly or implicitly claims that the case at hand is but an exemplification of certain general laws governing the development of neuroses. But surely, whatever specific laws of this kind might be adduced at present can purport, at the very best, to express probabilistic trends rather than deterministic uniformities: they may be construed as *laws of statistical form*, or briefly as *statistical laws*, to the effect that, given the childhood experiences in question—plus, presumably, certain particular environmental conditions in later life—there is such and such a statistical probability that a specified kind of neurosis will develop. Such statistical laws differ in form from strictly universal laws of the kind mentioned in our earlier examples of explanatory arguments. In the simplest case, a *law of strictly uni-*

---

planation. In this case, the laws invoked must meet certain conditions as to mathematical form, and C_1, C_2, . . . , C_n express so-called boundary conditions. For a fuller account of the concepts of causal law and of causality as understood in theoretical physics, see, for example, H. Margenau, *The Nature of Physical Reality* (New York: McGraw-Hill Book Company, Inc., 1950), Chapter 19; or Ph. Frank, *Philosophy of Science* (Englewood Cliffs, N.J.: Prentice-Hall, Inc., 1957), Chapters 11, 12.

versal form, or briefly, *a universal law,* is a statement to the effect that in *all* cases satisfying certain antecedent conditions *A* (e.g., heating of a gas under constant pressure), an event of a specified kind *B* (e.g., an increase in the volume of the gas) will occur; whereas a law of statistical form asserts that the probability for conditions *A* to be accompanied by an event of kind *B* has some specific value *p.*

Explanatory arguments which, in the manner just illustrated, account for a phenomenon by reference to statistical laws are not of the strictly deductive type (2.1). For example, the explanans consisting of information about Henry's exposure to the mumps and of a statistical law about the transmission of this disease does not logically imply the conclusion that Henry catches the mumps; it does not make that conclusion necessary, but, as we might say, more or less probable, depending upon the probability specified by the statistical laws. An argument of this kind, then, accounts for a phenomenon by showing that its occurrence is highly probable in view of certain particular facts and statistical laws specified in the explanans. An account of this type will be called an *explanation by inductive subsumption under statistical laws,* or briefly, an *inductive explanation.*

Closer analysis shows that inductive explanation differs from its deductive counterpart in several important respects; [3] but for the purposes of the following discussion, our sketchy account of explanation by statistical laws will suffice.

The two types of explanation we have distinguished will both be said to be varieties of *nomological explanation;* for either of them accounts for a given phenomenon by "subsuming it under laws," i.e., by showing that its occurrence could have been inferred—either deductively or with a high probability—by applying certain laws of universal or of statistical form to specified antecedent circumstances.

3. For details, see section 3 of the essay "Aspects of Scientific Explanation" in the volume, *Aspects of Scientific Explanation.* Some stimulating comments on explanation by means of statistical laws will be found in S. E. Gluck, "Do Statistical Laws Have Explanatory Efficacy?" *Philosophy of Science,* 22 (1955), 34–38. For a much fuller analysis of the logic of statistical inference, see R. B. Braithwaite, *Scientific Explanation* (Cambridge: Cambridge University Press, 1953), chapters V, VI, VII. For a study of the logic of inductive inference in general, Carnap's *Logical Foundations of Probability,* is of great importance.

Thus, a nomological explanation shows that we might in fact have *predicted* the phenomenon at hand, either deductively or with a high probability, if, at an earlier time, we had taken cognizance of the facts stated in the explanans.

But the predictive power of a nomological explanation goes much farther than this: precisely because its explanans contains general laws, it permits predictions concerning occurrences other than that referred to in the explanandum. In fact, such predictions provide a means of testing the empirical soundness of the explanans. For example, the laws invoked in a deductive explanation of the form (2.1) imply that the kind of event described in E will recur whenever and wherever circumstances of the kind described by C_1, C_2, . . . , C_n are realized; e.g., when the experiment with ice floating in water is repeated, the outcome will be the same. In addition, the laws will yield predictions as to what is going to happen under certain specifiable conditions which differ from those mentioned in C_1, C_2, . . . , C_n. For example, the laws invoked in our illustration also yield the prediction that if a chunk of ice were floating in a beaker filled to the brim with concentrated brine, which has a greater specific gravity than water, some of the liquid would overflow as the ice was melting. Again, the Newtonian laws of motion and of gravitation, which may be used to explain various aspects of planetary motion, have predictive consequences for a variety of totally different phenomena, such as free fall near the earth, the motion of a pendulum, the tides, and many others.

This kind of account of further phenomena which is made possible by a nomological explanation is not limited to future events; it may refer to the past as well. For example, given certain information about the present locations and velocities of the celestial bodies involved, the principles of Newtonian mechanics and of optics yield not only predictions about future solar and lunar eclipses, but also "postdictions," or "retrodictions," about past ones. Analogously, the statistical laws of radioactive decay, which can function in various kinds of predictions, also lend themselves to retrodictive use; for example, in the dating, by means of the radiocarbon method, of a bow or an ax handle found in an archaeological site.

A proposed explanation is scientifically acceptable only if its explanans is capable of empirical test, i.e., roughly speaking, if it is possible to infer from it certain statements whose truth can be checked by means of suitable observational or experimental procedures. The predictive and postdictive implications of the laws invoked in a nomological explanation clearly afford an opportunity for empirical tests; the more extensive and varied the set of implications that have been borne out by empirical investigation, the better established will be the explanatory principles in question.

Ernest Nagel

Ernest Nagel was born in Czechoslovakia and came to the United States as a boy. He has had a life-long association with Columbia University as a graduate student, as a professor, and as a leading representative of a philosophical tradition stemming from Dewey, Woodbridge, and Cohen. While influenced by pragmatism, logical positivism, and linguistic analysis, Nagel is too independent a thinker to be identified with any particular school. His work, *The Structure of Science,* from which the present excerpt is taken, almost immediately achieved the status of a classic in the philosophy of science. It is probably the most influential textbook in the field. The present selection, an evaluation of realism and operationalism as competing views of scientific explanation, illustrates the distinctive way in which Nagel relates detailed analyses of technical questions to the pragmatist's interest in what the scientist actually does with his scientific systems. Nagel is not concerned with differences that do not make a difference.

The Realist View of Theories

Are theories then 'really' statements of which truth and falsity are meaningfully predictable, despite the difficulties that have been noted in this view? Enough has already been said to suggest that, whether the question is answered affirmatively or negatively, the answer given may not be the exclusively reasonable one. Indeed, those who differ in their answers to it frequently disagree neither on matters falling into the province of experimental inquiry nor on points of formal logic nor on the facts of scientific procedure. What often divides them are, in part, loyalties to different intellectual traditions, in part inarbitrable preferences concerning the appropriate way of accommodating our language to the generally admitted facts. It is a matter of historical record that, while many distinguished figures in both science and philosophy have adopted as uniquely adequate the characterization of theories as true or false statements, a no less distinguished group of other scientists and philosophers has made a similar claim for the description of theories as instruments of inquiry. However, a defender of either view cannot only cite eminent authority to support his position; with a little dialectical ingenuity he can usually remove the sting from apparently grave objections to his position. In consequence, the already long controversy as to which of the two is the proper way of construing theories can be prolonged indefinitely. The obvious moral to be drawn from such a debate is that once both positions are so stated that each can meet the *prima facie* difficulties it faces, the question as to which of them is the "correct position" has only terminological interest.

From THE STRUCTURE OF SCIENCE, pp. 141–52, by Ernest Nagel, © 1961 by Harcourt, Brace & World, Inc. and reprinted with their permission and with the permission of Routledge & Kegan Paul Ltd., London.

1. Let us consider the chief obstacles to each of the two views under discussion, beginning with those facing the conception of theories as true or false statements.

a. There is in the first place the purely formal difficulty that a theory is not a statement, but only a statement-form. For if, as often happens, some terms of a theory are not associated with any correspondence rules, those terms are in effect variables, so that on the face of it the theory does not satisfy the grammatical requirements for statements. This difficulty can be met by a formal device, first explicitly proposed by Ramsey.[1] The device consists simply in introducing what are called 'existential quantifiers' as prefixes for statement-forms, so that the resulting expression will formally be a statement. For example, the expression "If a human being has a trait P, then such a person has blue eyes" is a statement-form; but by adding the prefix "There is a trait P," we obtain from it the statement "There is a trait P such that if a human being has P, then the person has blue eyes." Similarly, suppose that, although the terms "mass" and "acceleration" are associated with correspondence rules, the term "force" is not. The expression "If a body undergoes changes in motion, then the product of the mass and acceleration of the body is equal to the force F acting on it" is then in effect a statement-form, from which we can obtain the statement "If a body undergoes changes in motion, then there is a (measurable) property F such that the product of the mass and acceleration of the body is equal to F." More generally, let "$T(M,N,P,Q)$" be a theory whose theoretical terms "M" and "N" are associated with correspondence rules while its theoretical terms "P" and "Q" are not, so that "$T(M,N,P,Q)$" is by hypothesis a statement-form. Then "There is a P and there is a Q, such that $T(M,N,P,Q)$" is a statement. Accordingly, since by using the Ramsey device the observational consequences that can be derived from a theory are not altered, that device suffices for outflanking the formal difficulty under discussion.

b. In the second place, there is the objection previously mentioned that theories are commonly formulated in terms of limiting

1. Frank P. Ramsey, *op. cit.*, pp. 212–36.

concepts which characterize nothing actually in existence, so that at any rate nonvacuous factual truth cannot be claimed for such theories. This objection can be turned in a number of ways. A familiar gambit is to challenge the contention that limiting concepts do not apply to existing things. To be sure, we cannot, for example, ascertain by overt measurement the value of an instantaneous velocity of the magnitude of some length whose theoretical value is stipulated to be equal to the square root of 2. But unless accessibility to overt measurement (or more generally to observation) is made the criterion of physical existence, so it is sometimes said, this does not show that bodies cannot have instantaneous velocities or lengths with real number magnitudes. On the contrary, if a theory postulating such values is supported by competent evidence, then according to the rejoinder under discussion there is good reason to maintain that these limiting concepts do designate certain phases of things and processes. Since in testing a theory we test the totality of assumptions it makes, so the rejoinder continues, if a theory is regarded as well established on the available evidence, all its component assumptions must also be regarded. Accordingly, unless we introduce quite arbitrary distinctions, we cannot pick and choose between the component assumptions, counting some as descriptions of what exists and others as not.

There is another way in which the objection under discussion is sometimes countered. The rejoinder then consists in admitting that limiting concepts are simplifying devices, and that a theory employing them does not in general assert anything for which literal truth can reasonably be claimed. Nevertheless, existing things possess traits that often are either indistinguishable from the 'ideal' traits mentioned in a theory or differ from such 'ideal' traits by a negligible factor. In consequence, on this rejoinder to the objection, a theory is said to be true in the sense that the discrepancy between what a theory asserts and what even ultrarefined observation can discover is small enough to be counted as arising from experimental error.

c. A third type of difficulty for the conception of theories as true or false statements is created by the fact, to which attention has already been directed, that apparently incompatible theories are sometimes employed for the same subject matter. Thus, a liquid cannot be both a system of discrete particles and also a continuous medium, though theories dealing with the properties of liquids adopt one assumption in some cases and the opposing assumption in others.

The usual reply to this objection consists of two parts. One of them is essentially a repetition of the rejoinder mentioned in the preceding paragraph. A theory may be employed in a given area of inquiry, even though it is apparently incompatible with some other theory that is also used, because the former is simpler than the latter and because for the problems under discussion the more complex theory does not yield conclusions in better agreement with the facts than are the conclusions of the simpler theory. Accordingly, the simpler theory can be regarded as in a sense a special case of the more complex one, rather than as a contrary of the latter.

The second part of the reply is that, though incompatible theories may be used for a time, their use is but a temporary makeshift, to be abandoned as soon as an internally consistent theory is developed, more comprehensive than either of the previous ones. Thus, although there were serious discrepancies between the atomic theories employed at the turn of the present century to account for many facts of both physics and chemistry, these conflicting theories have been replaced by a single theory of atomic structure currently although there were serious discrepancies between the atomic theories, each of which is nevertheless useful in some limited domain of inquiry, are often a powerful incentive for the construction of a more inclusive but consistent theoretical structure. Accordingly, a proponent of the view that theories are true or false statements can escape any embarrassment for his position from the circumstance that incompatible theories are sometimes employed in the sciences; he can insist on the corrigible character of every theory and refuse to claim final truth for any theory. He can freely admit that even a false theory may be quite useful for handling many problems; and he can join this admission with the claim that the succession of theories in some branch of science is a series of progressively better approximations to the unattainable but valid ideal of a finally true theory.

d. And finally, there is the objection currently raised against the position under discussion because of the difficulties encountered in interpreting quantum mechanics in terms of some familiar model. For example, theoretical as well as experimental considerations have led physicists to ascribe to electrons (and to other entities postulated by quantum theory) apparently incompatible and in any case puzzling characteristics. Thus, electrons are construed to have features which make it appropriate to think of them as a system of waves; on the other hand, electrons also have traits which lead us to think of them as particles, each having a spatial location and a velocity, though no determinate position and velocity can in principle be assigned simultaneously to any of them. Many physicists have therefore concluded that quantum theory cannot be viewed as a statement about an 'objectively existing' domain of things and processes, as a map that outlines even approximately the microscopic constitution of matter. On the contrary, the theory must be regarded simply as a conceptual schema or a policy for guiding and coordinating experiments.

The rejoinder to this objection follows a familiar pattern. The fact that a visualizable model embodying the laws of classical physics cannot be given for quantum theory, so runs the reply, is not an adequate ground for denying that the quantum theory does formulate the structural properties of subatomic processes. It is doubtless desirable to have a satisfactory model for the theory. But the type of model that is regarded as satisfactory at any given time is a function of the prevailing intellectual climate. Even though current models for quantum theory may strike us as strange and even 'unintelligible', there are no compelling reasons for assuming that the strangeness will not wear away with increased familiarity, or that a more satisfactory interpretation for the theory will not be eventually found. Moreover, the alleged unintelligibility of the present model stems in large measure from a failure to note that words like "wave" and "particle" used in describing it are being employed in an analogical manner. It is only in a Pickwickian sense that an electron is a particle (in the customary meaning of the word), just as it is in a stretched sense that $\sqrt{-1}$ is a number (in the sense in which the cardinal integer 3 is a number). An electron is said to be a particle (or alternatively, a wave) because some of the properties ascribed to electrons are analogous to certain properties associated with classical particles, or alternatively, with familiar water waves, even if the analogy fails for other properties.

When the language of "particles" and "waves" is understood in terms of the way these words are actually used in the context of quantum mechanics, so it has been contended, not even the appearance of contradiction arises in the quantum-theoretical characterizations of electrons. But in any event, the basic issue is not whether a particular substantive model of subatomic processes is satisfactory. The basic issue is whether the relations between elementary constituents of physical objects and processes are more adequately stated by the mathematical formalism of quantum mechanics than by any other formal model available at present. On this issue, there is no disagreement among competent students that the answer is affirmative.

This sample of objections to the view that theories are true or false statements suffices to show that the view has dialectical resources for maintaining itself in the face of severe criticism. Undoubtedly the rejoinders to these criticisms can be met with counter-rejoinders, though none to which defenders of the view under attack cannot offer at least a *prima facie* suitable reply. It would therefore not be profitable to continue this phase of the discussion any further. Let us turn instead to some of the criticisms of the instrumentalist position.

2. Two main difficulties have been noted in the instrumentalist position as usually formulated. The first of them is that much experimental research is directed to finding evidence for or against a theory—an undertaking which is apparently pointless if a theory is not a genuine statement but simply a formulation of policy or rule of procedure. However, this objection can be readily made innocuous. For it is sufficient to reply that a theory can indeed be 'tested' by searching for evidence which will either 'confirm' or 'refute' it, but only in the sense that confirmatory or disconfirmatory evidence is sought for observational conclusions drawn from observational premises in accordance with the theory. As we have seen, the sole issue raised by this way of putting the matter concerns the relative convenience of employing *material* rather than *purely formal* leading principles in reconstructing deductive inferences.

The second and more serious difficulty is that a consistently held instrumentalist view apparently precludes its adherents from admitting the 'physical reality' (or 'physical existence') of any 'scientific objects' ostensibly postulated by a theory. For if a theory employing such terms as "atom" or "electron" is merely a leading principle, it is incongruous to ask whether there 'really are' atoms; and it is then acutely puzzling to say, as some physicists do, that because of the experimental evidence now 'pointing' to the atom, "we are as convinced of its physical existence as of our hands and feet."

However, the force of this objection is unclear because of the notorious ambiguity if not obscurity of the expression 'physical reality' or 'physical existence'. In any event, writers using these phrases do not in general understand them in the same sense. It will therefore be useful to consider some of the different criteria that are commonly employed, whether explicitly or tacitly, when physical reality is either affirmed or denied of scientific objects such as electrons, atoms, electric fields, and the like.

a. For anything to be regarded as physically real, perhaps the most familiar requirement is that the thing or event be publicly perceived when suitable conditions for its observation are realized. In terms of this criterion, sticks, stones, flashes of lightning, the smells of cooking, and the like can be said to exist physically, but not the pains a man feels when he turns an ankle, nor the pink elephants a drunkard may experience in his delirium. However, most scientific objects are not physically real in this sense. Thus, although illuminated surfaces are physically real on this criterion, light waves are not; and, although the condensations of water vapor to form visible tracks in a Wilson cloud chamber are real, the alpha particles which (according to current physical theory) produce those tracks are not. Certainly it is not on this interpretation of 'physically real' that we are as convinced of the physical reality of atoms as we are of our hands and feet. On the other hand, even if some hypothetical scientific objects were physically real in this sense —for example, if the genes postulated by current biological theory of heredity could be made visible—the role of the theoretical notions in science, in terms of which such objects are specified, would not be altered. It is of course quite possible that if we could perceive molecules, many questions still outstanding about them would be answered, so that molecular theory would receive an improved formulation. Nevertheless, molecular theory would still continue to formulate the traits of molecules in *relational* terms—in terms of relations of molecules to other molecules and to other things—not in terms of any of their qualities that might be directly apprehended through our organs of sense. For the *raison d'être* of molecular theory is not to supply information about the sensory qualities of molecules but to enable us to understand (and predict) the occurrence of events and the relations of their interdependence in terms of pervasive structural patterns into which they enter. Accordingly, in this sense of the phrase the physical reality of theoretical entities is of little import for science.

b. A second, widely accepted criterion of physical reality is close to being the polar opposite of the first, and has already been mentioned in passing. According to it, every nonlogical term of an assumed law (whether experimental or theoretical) designates something that is physically real, provided that the law is well supported by empirical evidence and is generally accepted by the scientific community as likely to be true. On this criterion, therefore, physical reality is ascribed not only to such experimentally identifiable items as the kinetic energy of a bullet, the strain in a body subjected to stresses, the viscosity of a liquid, or the electrical resistance of a wire but also to theoretical objects like light waves, atoms, neutrinos, and waves of probability. Anyone who employs this criterion will accordingly hold that many objects postulated by some accepted theory are physically existing things, even before any empirical evidence confirming detailed specific assumptions about those objects is available. This seems to have been the criterion adopted by many contemporary physicists who believed in the physical existence of antiprotons as postulated by quantum theory, although definite experimental evidence for them was lacking until recently. On the other hand, those who employ this criterion will deny physical reality to a scientific object once so characterized (such as the phlogiston substance postulated by the phlogiston theory of combustion) when the theory postulating that object is abandoned as unsatisfactory—unless, indeed, a different but acceptable theory postulates a closely analogous object.

c. A third criterion of physical reality sometimes employed is that a term designating anything physically real must enter into more than one experimental law, with the proviso that the laws are logi cally independent of each other and that none of them is logically equivalent to a set of two or more laws. This requirement can obviously be strengthened by demanding that there be a considerable number of such experimental laws. The rationale for this requirement is to characterize as physically real only things that can be identified in ways other than, and independently of, the procedures used to define those things. For example, the magnitude of the gravitational force of the earth on a body appears as the constant "g" in Galileo's law for freely falling bodies. If this were the only law in which "g" occurred, then on this criterion the term "gravitational force" would not designate a physical reality. However, "g" enters into a number of other experimental laws, such as the law for the period of the simple pendulum. Accordingly, physical reality can be ascribed to the gravitational force of the earth. On the other hand, the situation appears to be different in the case of the notion of electric field. We can determine the strength of an electric field in a region by introducing into that region a test body with known mass and electric charge, and measuring the force upon that body. The field strength is then defined as the ratio of the force to the charge on the body; and it is an experimental law that under specified conditions this ratio has the same constant value for any test body of relatively small dimensions. Though in this way the term "electric field" enters into an experimental law, this seems to be the only experimental law in which the term does occur. If so, then according to the present criterion physical reality cannot be ascribed to electric fields.

The application of this criterion to scientific objects postulated by microscopic theories involves some complications, since theoretical terms do not occur in statements of experimental laws. It would take us too far afield to unravel these complications in any detail. In any event, it will suffice for present purposes to construe the criterion for the physical reality of theoretical entities as saying that the theoretical term ostensibly referring to such entities must be *associated* by correspondence rules with experimental concepts, and furthermore that these experimental concepts must enter into at least two logically independent experimental laws which can be derived from the theory. For example, in the kinetic theory of gases theoretical expressions such as the "mass of a molecule," the "mean kinetic energy of molecules," the "number of molecules," and the like are associated with experimental concepts such as the "mass of a gas," the "temperature of a gas," and the "ratio of the product of the pressure and volume of a gas to its temperature." These latter terms occur in several experimental laws, such as the Boyle-Charles' law, Dalton's law of partial pressures, or the law that at given temperature and pressure the difference of the two specific heats per unit volume is the same for all gases—all of which are laws derivable from the theory.

It is worth noting that on this criterion of physical existence, not every entity postulated by a theory will in general be said to exist, even if the theory as a whole is well confirmed by experiment and is accepted as likely to be true. Thus, some physicists were once doubtful about the physical existence of neutrinos, initially postulated to preserve the conservation of energy principle in quantum theory; and it is possible that this doubt was based on the fact that the term "neutrino" did not conform to the requirement set by this criterion. Similarly, when Planck first introduced the theoretical notion of discrete quanta of energy in order to account successfully for the distribution of energy in the spectrum of black-body radiation, physicists (including Planck himself) were doubtful of the existence of such quanta. The situation was altered when the notion of energy quanta was associated with the constant "h" that appeared not only in Planck's radiation law but also in other experimental laws concerning the photoelectric effect, the line spectra of the elements, the specific heats of solids, and so on, all of which were derived from theories containing the quantum hypothesis as one component assumption.

d. A fourth, and in some ways more restrictive, criterion of physical reality is often adopted. On this criterion, a term signifies something physically real, if the term occurs in a well-established "causal law" (whether theoretical or experimental), in some indicated sense of 'causal'. In a more specific version of this criterion, the term must describe what is technically called the 'state of a physical system', so that if "A_t" is the state-description of a system at time t, the causal law asserts that the given state is invariably followed (or preceded) by the state $A_{t'}$ at time t' later (or earlier) than t.[2]

2. The notion of 'state' will be discussed more fully in the next chapter [of original work. Ed.]

For example in mechanics the state of a system of particles is described by the set of numbers that specify the positions and velocities of the particles. The causal laws of mechanics enable us, given the positions and velocities of a set of particles at any initial time, to determine their positions and velocities at any other time. Accordingly, the mechanical state of a system is physically real. Similarly, the state of a system in quantum theory is described by a certain function (known as the Psi-function) of positions and energies of elementary particles, where the function is a solution of the fundamental wave equation of the theory. This equation in effect asserts that the Psi-state of a system at some given time is invariably succeeded by the calculable Psi-state of the system at any specified future time. Accordingly, on the present criterion, the Psi-state is physically real. On the other hand, since in quantum mechanics the coordinates of position and velocity of an individual elementary particle, such as an electron, do not constitute the state-description of the particle, they do not describe what is physically real. In the view of at least some physicists, physical reality cannot therefore be ascribed to individual electrons and other such atomic entities.[3]

e. One final criterion of physical reality is worth noting, according to which the real is that which is invariant under some stipulated set of transformations, changes, projections, or perspectives. An elementary geometrical example will illustrate the general idea underlying this criterion. Imagine a circle painted on a sheet of glass on a horizontal plane and a small source of light at some distance perpendicularly above the center of the circle. The circle will then be projected as a shadow cast on a screen parallel to the glass, and this shadow will also be a circle. Suppose, however, that the glass is rotated through an axis passing through the glass and parallel to the screen, with the source of light and the screen remaining in their initial positions. The shadows on the screen will then no longer be circles; they will first assume the form of ellipses, and eventually take the form of parabolas. Under this

3. Cf. the discussion of this point in the debate between two leading contemporary physicists, Erwin Schrödinger, "Are There Quantum Jumps?" *British Journal for the Philosophy of Science,* Vol. 3 (1952), pp. 109–23, 233–42; and Max Born, "The Interpretation of Quantum Mechanics," *British Journal for the Philosophy of Sceince,* Vol. 4 (1953), pp. 95–106.

projection, neither the shape nor the perimeter nor the area of the circle on the glass will be preserved in the circle's shadow: these are not invariant properties of the circle under projection. Nevertheless, some properties of the circle are invariant under such projection. For example, if a straight line is painted on the glass to intersect the circle, the shadow of the line will always intersect the shadow of the circle in two points. If the present criterion were applied to this example, we would have to say that neither the shape nor the perimeter nor the area of the figure on the glass is a physical reality, but that only properties of the figure invariant under projection (such as the one mentioned) are physically real.

It will be evident that on this criterion different sorts of things can be characterized as physical realities, according to which set of transformations is specified for this purpose. Thus, some writers have denied physical reality to immediate sensory qualities, since these vary with physical, physiological, and even psychological conditions. The title to such reality has been reserved by these thinkers for the so-called "primary qualities" of things, whose interrelations are independent of physiological and psychological changes and are formulated by the laws of physics. Similarly, the numerical value of the velocity of a body is not invariant when the motion of the body is referred to different frames of reference, so that on this criterion relative velocity is not a physical reality. Many writers on the theory of relativity have in fact maintained that spatial distances and temporal durations as conceived by pre-relativity physics are not physically real, since they are not invariant for all systems moving with respect to each other with constant relative velocities. Physical reality, according to these writers, must be ascribed only to those features of things that are formulated by the invariant laws of relativity physics (such as the relativistic kinetic energy of a body or its relativistic momentum). In an analogous manner, physical reality has been attributed to theoretical entities like atoms, electrons, mesons, probability waves, and the like because they satisfy some indicated condition of invariance.

To prevent possible misunderstanding, it is perhaps worth stressing that the criteria mentioned in the preceding discussion are intended to be explicative of what is supposedly meant in a number of contexts when something is said to be physically real. The ascription of physical reality in any of the senses distinguished must therefore not be misconstrued. It must not be understood as implying that a thing so characterized has a place in the scheme of things to be contrasted with certain other things having the invidious label of 'mere appearance', or that in addition to satisfying the requirements specified by the corresponding criterion the thing is in some way more valuable or more fundamental than everything not so characterized. Many scientists as well as philosophers have indeed often used the term 'real' in an honorific way to express a value judgment and to attribute a "superior" status to the things asserted to be real. There is perhaps an aura of such honorific connotation whenever the word is employed, despite explicit avowals to the contrary and certainly to the detriment of clarity. For this reason it would be desirable to ban the use of the word altogether. As things stand, however, linguistic habits are too deeply ingrained and too widespread to make such a ban possible. Accordingly, these cautionary remarks have been added in order to make clear that any invidious contrasts which may be suggested by the word "real" are irrelevant to the present discussion.

In any event, this brief list of criteria does not exhaust the senses of 'real' or 'exist' that can be distinguished in discussions about the reality of scientific objects. It is long enough, however, to indicate that a proponent of the instrumentalist view of theories cannot give an unambiguous answer to the ambiguous question whether it is congruous with his position to accept the physical reality of such things as atoms and electrons. But the list is also long enough to suggest that there are at least some senses of the expressions 'physically real' and 'physically exist' in which an ironically minded instrumentalist can acknowledge the physical reality or existence of many theoretical entities.

More specifically, if the third of the above criteria is adopted for specifying the sense of 'physically real', it is quite patent that the instrumentalist view is entirely compatible with the claim that atoms, say, are indeed physically real. In point of fact, many instrumentalists themselves urge such a claim. To make the claim is to assert that there are a number of well-established experimental laws related in a certain manner to one another and to other laws by way of a given atomic theory. In short, to assert that in this sense atoms exist is to claim that available empirical evidence is sufficient to establish the adequacy of the theory as a leading principle for an extensive domain of inquiry. But as has already been noted, this is in effect only verbally different from saying that the theory is so well confirmed by the evidence that the theory can be tentatively accepted as true.

Proponents of the instrumentalist position may, of course, reserve judgment about whether other theoretical entities postulated by the theory really do exist, since the requirements for their physical reality as set by the criterion adopted may not be clearly satisfied. But on such particular issues proponents of the view that theories are true or false statements may have similar hesitations. It is therefore difficult to escape the conclusion that when the two apparently opposing views on the cognitive status of theories are each stated with some circumspection, each can assimilate into its formulations not only the facts concerning the primary subject matter explored by experimental inquiry but also all the relevant facts concerning the logic and procedure of science. In brief, the opposition between these views is a conflict over preferred modes of speech.

Willard Van Orman Quine

Willard Van Orman Quine was born in Akron, Ohio, and studied logic under Alfred North Whitehead at Harvard. One of Quine's earliest works, *New Foundations for Mathematical Logic,* quickly established him as one of America's leading logicians. In addition to his technical contributions to logic, Quine has written extensively on the foundations of mathematics, on the relation of logic to linguistic analysis, and on epistemological problems. One of his main concerns is a systematic unification of knowledge through the use of an adequate logical formalism. The present selection outlines his views later developed in much greater detail in *Word & Object.* As in this simple summary, his later views are characterized by a maximal reliance on formal logic, a minimal reliance on—though never an outright dismissal of—ontology, and a frank acceptance of the role that personal decisions play in the implementation of any system. In this way Quine developed a philosophy that represents a synthesis of elements drawn from pragmatism, positivism, and linguistic analysis. Since 1955 Quine has been a professor of philosophy at Harvard.

On the Role of Logic in Explanation

Logic, like any science, has as its business the pursuit of truth. What are true are certain statements; and the pursuit of truth is the endeavor to sort out the true statements from the others, which are false.

Truths are as plentiful as falsehoods, since each falsehood admits of a negation which is true. But scientific activity is not the indiscriminate amassing of truths; science is selective and seeks the truths that count for most, either in point of intrinsic interest or as instruments for coping with the world.

For truth ordinarily attaches to statements by virtue of the nature of the world. It is a commonplace, inaccurate but not unfounded, that a statement is true when it corresponds to reality, when it mirrors the world. A fundamental way of deciding whether a statement is true is by comparing it, in some sense or other, with the world—or, which is the nearest we can come, by comparing it with our experience of the world.

Strictly speaking, what admit of truth and falsity are not statements as repeatable patterns of utterance, but individual events of statement utterance. For, utterances that sound alike can vary in meaning with the occasion of the utterance. This is due not only to careless ambiguities, but to systematic ambiguities which are essential to the nature of language. The pronoun "I" changes its reference with every change of speaker; "here" changes its reference with every significant movement through space; and "now" changes its reference every time it is uttered.

So the crucial point of contact between description and reality is

From the introduction, pp. xi–xvii, to METHODS OF LOGIC, Revised Edition, by Willard Van Orman Quine. Copyright 1950, © 1959 by Holt, Rinehart and Winston, Inc. Reprinted by permission of Holt, Rinehart and Winston, Inc.

151

to be sought in the utterance of a statement on the occasion of an experience which that statement utterance directly reports. The seeing of a green patch, and the simultaneous utterance "Green patch now," constitute the sort of composite event which, in its rare occurrences, gladdens the heart of the epistemologist.

Such events, fundamental though they are epistemologically, are rare because of the social nature of language. Language is a social institution serving, within its limitations, the social end of communication; so it is not to be wondered that the objects of our first and commonest utterances are socially shared physical objects rather than private experiences. Physical objects, if they did not exist, would (to transplant Voltaire's epigram) have had to be invented. They are indispensable as the public common denominators of private sense experience.

But utterances about physical objects are not verifiable or refutable by direct comparison with experience. They purport to describe, not experience, but the external world. They can be compared with the external world only through the medium of our experience of that world, but the connection between our experience and the world already involves a step of hypothesis or inference which precludes any direct and conclusive confrontation of the utterance with its subject matter. There is many a slip betwixt objective cup and subjective lip.

So statements, apart from an occasional collectors' item for epistemologists, are connected only deviously with experience. The latest scientific pronouncement about positrons and the statement that my pen is in my hand are equally statements about physical objects; and physical objects are known to us only as parts of a systematic conceptual structure which, taken as a whole, impinges at its edges upon experience. As far as knowledge is concerned, no more can be claimed for our whole body of affirmations than that it is a devious but convenient system for relating experiences to experiences. The system as a whole is under-determind by experience, but implies, given certain experiences, that certain others should be forthcoming. When such predictions of experience turn out wrong, the system has to be changed somehow. But we retain a wide latitude of choice as to what statements of the system to preserve and what ones to revise; any one of many revisions will be sufficient to unmake the particular implication which brought the system to grief. Our statements about external reality face the tribunal of sense experience not individually but as a corporate body.

But such choice of what to revise is subject to a vague scheme of priorities. Some statements about physical objects, e.g., "My pen is in my hand," "The mercury is at 80," are in some sense closer to possible experience than others; and such statements must be guarded pretty jealously once the appropriate experiences have appeared. Should revision of the system become necessary, other statements than these are to suffer. It is only by such an allocation of priority that we can hope to claim any empirical content or objective reference for the system as a whole.

There is also, however, another and somewhat opposite priority: the more fundamental a law is to our conceptual scheme, the less likely we are to choose it for revision. When some revision of our system of statements is called for, we prefer, other things being equal, a revision which disturbs the system least. Actually, despite the apparent opposition between this priority and the one previously noted, the one involves the other. For, the connection between a statement such as "My pen is in my hand" and the experiences which are said to verify it is itself a matter of general principles central to the system.

Where the two priorities come into conflict, either is capable of prevailing. Statements close to experience and seemingly verified by the appropriate experiences may occasionally be given up, even by pleading hallucination, in the extreme case where their retention would entail a cataclysmic revision of fundamental laws. But to overrule a multiplicity of such statements, if they reinforce one another and are sustained by different observers, would invite criticism.

The priority on law, considered now apart from any competition with the priority on statements verified by experience, admits of many gradations. Conjectures of history and economics will be revised more willingly than laws of physics, and these more willingly than laws of mathematics and logic. Our system of statements has such a thick cushion of indeterminacy, in relation to experience, that vast domains of law can easily be held immune to revision on principle. We can always turn to other quarters of the system when revisions are called for by unexpected experiences. Mathematics and logic, central as they are to the conceptual scheme, tend to be accorded such immunity, in view of our conservative preference for revisions which disturb the system least; and herein, perhaps, lies the "necessity" which the laws of mathematics and logic are felt to enjoy.

In the end it is perhaps the same to say, as one often does, that the laws of mathematics and logic are true simply by virtue of our conceptual scheme. For, it is certainly by virtue of that scheme that those laws are central to it; and it is by virtue of being thus central that the laws are preserved from revision at the expense of statements less strategically situated.

It is also often said that the laws of mathematics and logic are true by virtue of the meanings of the words "+," "=," "if," "and," etc., which they contain. This also I can accept, for I expect it differs only in wording from saying that the laws are true by virtue of our conceptual scheme.

But it must now be remarked that our conservative preference for those revisions which disturb the system least is opposed by a significant contrary force, a force for simplification. Far-reaching revision of the fundamental laws of physics was elected in recent decades, by considerations of simplicity, in preference to the welter of *ad hoc* subsidiary laws which would otherwise have been needed to accommodate the wayward experiences of Michelson and Morley and other experimenters. Continued experiment 'confirmed' the fundamental revisions, in the sense of increasing the simplicity differential.

Mathematical and logical laws themselves are not immune to revision if it is found that essential simplifications of our whole conceptual scheme will ensue. There have been suggestions, stimulated largely by quandaries of modern physics, that we revise the true-false dichotomy of current logic in favor of some sort of tri- or *n*-chotomy. Logical laws are the most central and crucial statements of our conceptual scheme, and for this reason the most protected from revision by the force of conservatism; but, because again of their crucial position, they are the laws an apt revision of which might offer the most sweeping simplification of our whole system of knowledge.

Thus the laws of mathematics and logic may, despite all 'necessity' be abrogated. But this is not to deny that such laws are true by virtue of the conceptual scheme, or by virtue of meanings. Because these laws are so central, any revision of them is felt to be the adoption of a new conceptual scheme, the imposition of new meanings on old words. No such revolution, by the way, is envisaged in this book; there will be novelties of approach and technique in these pages, but at bottom logic will remain unchanged.

For the most part, as has been stressed in the foregoing paragraphs, our statements relate only remotely to experience. The system of statements as a whole has its experiential implications; but the individual statements, apart from the peripheral few which directly describe experience as such, are relevant to experience only indirectly through their participation in the system. It is only by way of the relations of one statement to another that the statements in the interior of the system can figure at all in the prediction of experience, and can be found deserving of revision when prediction fails. Now of these relations of statements to statements, one of conspicuous importance is the relation of logical implication: the relation of any statement to any that follows logically from it. If one statement is to be held as true, each statement implied by it must also be held as true; and thus it is that statements internal to the system have their effects on statements at the periphery.

But for implication, our system of statements would for the most part be meaningless; nothing but the periphery would make sense. Yet implication is not really an added factor; for, to say that one statement logically implies a second is the same as saying that a third statement of the system, an "if-then" compound formed from the other two, is logically true or 'valid'. Logical truths are statements on a par with the rest, but very centrally situated; they are statements of such forms as "$x = x$," "p or not p," "If p then p," "If p and q then q," "If everything is thus and so then something is thus and so," and others more complex and less quickly recognizable. Their characteristic is that they not only are true but stay true even when we make substitutions upon their component words and phrases as we please, provided merely that the so-called "logical" words "$=$," "or," "not," "if-then," "everything," "something," etc., stay undisturbed. We may write any statements in the "p" and "q" positions and any terms in the "thus and so" positions, in the forms cited above, without fear of falsity. All that counts, when a state-

ment is logically true, is its structure in terms of logical words. Thus
it is that logical truths are commonly said to be true by virtue merely
of the meanings of the logical words.

The chief importance of logic lies in implication, which, therefore,
will be the main theme of this book. Techniques are wanted for
showing, given two statements, that the one implies the other; herein
lies logical deduction. Such techniques will be developed, for increasingly inclusive portions of logic, as the book proceeds. The
objects of deduction, the things related by implication, are statements; so statements will constitute not merely the medium of this
book (as of most), but the primary subject matter.

Strictly speaking, as urged earlier, what admit of meaning and of
truth and falsity are not the statements but the individual events of
their utterance. However, it is a great source of simplification in
logical theory to talk of statements in abstraction from the individual
occasions of their utterance; and this abstraction, if made in full
awareness and subject to a certain precaution, offers no difficulty.
The precaution is merely that we must not apply our logical techniques to examples in which one and the same statement recurs
several times with changed meanings, due to variations in immediate context. But such examples are easily enough adjusted to the
purposes of logic by some preliminary paraphrasing, by way of
bringing the implicit shifts of meaning into explicit form.

Logic and mathematics were coupled, in earlier remarks, as jointly
enjoying a central position within the total system of discourse. Logic
as commonly presented, and in particular as it will be presented in
this book, seems to differ from mathematics in that in logic we talk
about statements and their interrelationships, notably implication,
whereas in mathematics we talk about abstract nonlinguistic things:
numbers, functions, and the like. This contrast is in large part misleading. Logical truths, e.g., statements of the form "If p and q then
q," are not about statements; they may be about anything, depending on what statements we put in the blanks "p" and "q." When we
talk *about* such logical truths, and when we expound implications,
we are indeed talking about statements; but so are we when we talk
about mathematical truths.

But it is indeed the case that the truths of mathematics treat explicitly of abstract nonlinguistic things, e.g., numbers and functions,
whereas the truths of logic, in a reasonably limited sense of the word

"logic," have no such entities as specific subject matter. This is an important difference. Despite this difference, however, logic in its higher reaches is found to bring us by natural stages into mathematics. For, it happens that certain unobtrusive extensions of logical theory carry us into a realm, commonly also called "logic" in a broad sense of the word, which does have abstract entities of a special kind as subject matter. These entities are classes; and the logical theory of classes, or set theory, proves to be the basic discipline of pure mathematics. From it, as first came to be known through the work of Frege, Dedekind, Weierstrass, and their successors within the past seventy years, the whole of classical mathematics can be generated. Before the end of the book we shall have ascended through four grades of logic in the narrower sense, and emerged into set theory; and here we shall see, as examples of the derivation of classical mathematics, how the concept of number and various related notions can be defined.

Errol E. Harris

Errol E. Harris was born in South Africa and educated in South Africa and at Oxford University. Since 1956 he has been teaching philosophy in the United States and is currently a professor of Philosophy at Northwestern University. His recent work, *The Foundations of Metaphysics in Science*, constitutes a distinct alternative to empiricists' interpretations of scientific knowledge. The included summary of his epistemological views was presented as an invited paper to the *American Catholic Philosophical Association*. It is interesting to contrast his views with Quine's. Both men seek a unified systematization of knowledge which leads them to stress coherence and overall simplicity. Both accord scientific knowledge a primary role and attempt to make explicit the metaphysics implicit in the foundations of scientific knowledge. Yet, differing epistemological orientations lead to radically different philosophical positions. For Harris, the ultimate ground for a systematic coherence is the way the universe is. Hence, he constantly seeks to show the rationality of reality. For Quine, a rational coherence is imposed on reality by the conceptual systems we construct to interpret the incessant barrage of stimuli that affects our senses. This contrast brings out the decisive role that epistemological considerations play in any development of, or attack on, a doctrine of scientific realism.

Method and Explanation in Metaphysics

Positivism in its heyday in the 1930's and 40's restricted the methods of attaining knowledge and of expressing the results of research to those of the natural sciences. By 'knowledge', here, is meant all that is of primary importance, that is, knowledge about the world. The positivists allowed a secondary and derivative kind of knowledge, about the symbolism and language used for the expression of the primary sort, which they made the province of philosophy, confining it to semiotic—syntactics, semantics and the like. But for metaphysics no place was found, no subject matter, no legitimate or significant mode of expression, no method of procedure.

In the last two decades positivism has steadily declined. The verification principle, on which the rejection of metaphysics rested, was found wanting and succumbed to criticism, both from within the empiricist camp and from without. But its influence is still powerful and its implications are still accepted, if only tacitly, by an influential body of contemporary writers. Though metaphysics has again been permitted some place among theoretical disciplines and has been readmitted to the realm of sensible discourse, it is still with doubt and hesitancy that it is mentioned, and the general run of professional philosophers acknowledge no great desire to pursue it. Though it is no longer castigated as disreputable, much question is raised, and little agreement is reached, as to its proper method and actual province. P. F. Strawson has made concessions to it by distinguishing what he has called descriptive from revisionary metaphysics, conferring upon the former a safe respectability by assimilating it, in large measure, to linguistic analysis (particularly to ordinary-language philosophy). But the 'revisionary' variety is treated with more caution or is avoided altogether. In spite of Strawson's

From the *Proceedings of The American Catholic Philosophical Association*, 41: (1967): 124–33, by permission of the editor and the author.

venture, many still associate metaphysics with vagueness and mystification; and others regard it at best (though with obvious inconsistency) as the fabrication of hypotheses for which no empirical evidence is obtainable, which, by systematic reinterpretation of primitive terms and prescription of theoretical concepts, makes empirical refutation of these hypotheses impossible, yet which is valuable as a foil and a source of alternatives to current scientific theories (like quantum theory) which present special philosophical difficulties.[1]

But as the verification principle has collapsed in well-nigh all its recognizable forms, the ban against metaphysics should be altogether void and we should be able to proceed as was done before it was pronounced. One of its most lasting effects, however, has been to discredit the methods and theories of older philosophies and to raise a generation of students partly ignorant of the older tradition and partly biased by ready-made criticism so that what was done before is both unfamiliar and suspect. If we are to rediscover an appropriate method of metaphysics, therefore, we should not be ill-advised to glance once again at the historical record.

By this I do not mean, so much, that we should remind ourselves what Plato, or Aristotle, or Spinoza, said was the correct method of speculation, but rather what was the practice of these and other great philosophers and its relation to other sciences. We are all aware that the very earliest Greek philosophers were both metaphysicians and natural scientists at once; not that they all pursued both disciplines as separate branches of study, but that for them no distinction was drawn between science and philosophy and none is imputed to them by later commentators. This state of affairs persisted until Aristotle subdivided the forms of theoretical knowledge into different sciences, but even he did not regard metaphysics as really divorceable from physics, the connexion being so close that, included in the treatise which gives the science its name, is a portion of the Physics itself. Right up to the 18th century, the term 'philosophy' continued to cover all the exact and the natural sciences, as well as everything that goes by that name today, and even now in the Uni-

1. Cf., P. Feyerabend, "Problems of Empiricism." *Beyond the Edge of Certainty,* R. Colodny, Ed. (Englewood Cliffs, N.J.: Prentice-Hall, 1965), pp. 178 and 181 ff.

versity of Cambridge and in the old Scottish Universities physics is still known as 'natural philosophy'.

Moreover, many of the great metaphysicians of the modern period, Descartes, Leibniz, Kant, Bergson and Whitehead were also scientists, and no considerable philosophical theory has been uninfluenced by scientific discovery or failed in its turn to have some effect on the course of scientific research. In the words of A. E. Taylor "Every great metaphysical conception has exercised its influence on the general history of science, and, in return, every movement in science has affected the development of metaphysics. . . .[2] This persistent historical association between science and metaphysics cannot be without significance and is symptomatic of an intimate relation between both their methods and their subject matter.

Curiously enough, the positivistic indictment against metaphysics is that it lacks what is taken to be the hallmark of science—empirical reference—and that its method is one of speculation and *a priori* deduction, as opposed to induction and verification by empirical evidence. I shall attempt to defend the accused against these charges by showing that they misconceive the methods of science, on the one hand, and misrepresent that of metaphysics, on the other, and that the long historical association between metaphysics and the natural sciences is normal because their methods are in principle the same and their subject matters akin. I shall maintain that, so far from neglecting empirical evidence, metaphysical theory rests upon it just to the extent that it is provided and interpreted by the natural sciences, and that the relation between theory and observation is the same for both.

I shall begin with the latter.

Observation is the deliverance of sense-perception, and all the current and most widely canvassed theories of perception today seem to agree in rejecting the notion that the percept is an immediately given atomic and simple datum isolable from all others and from other elements in experience. Ryle's view that it is an achievement has come to be widely accepted among analysts, who point out that all verbs of perceiving, as used in reference to acts of perception, are achievement words. They imply that some piece of in-

2. A. E. Taylor, *Elements of Metaphysics* (New York: Macmillan, 1916), p. 13.

formation has been attained, presumably as a result of some effort or process of trial. Such verbs are held to have intentionality and imply the presence of an actual (material) object, and a knowledge of its presence. No such knowledge could be given as a simple unmediated datum, for the presence of a physical object is a complex fact, involving its solidity and persistence as well as its causal efficacy and liability to extraneous causal influences. Without these, no material object would actually be presented. The achievement (for instance) of seeing such an object must thus be the end-result of a discursive process of some kind, however rapidly accomplished, and could not be the result of any merely immediate intuition of an unprocessed datum.

The old idea of the existence and apprehension by the senses of sense-data has long been scouted, and philosophers today are more apt to regard material objects, in all their complexity of character and mutual relation, as the primary objects of perception. If this is not precisely what they say, it is the direct implication of the views set out at length by P. F. Strawson in *Individuals,* Stewart Hampshire in *Thought and Action* and D. M. Armstrong in *Perception and the Physical World.* It is also implied in Alan Donagan's critique of Collingwood.[3]

In Armstrong's terms, an act of perception is an act of acquiring knowledge, or belief, about particular facts in the physical world by the use of the senses. But no knowledge of a particular fact in the physical world is a simple matter. Take Armstrong's favorite example—that the cat is on the mat. To know (or believe) this involves knowing what sort of things cats and mats are, and how they are distinguished from other, different, things. It involves an understanding of spatial relations in general and that of 'being on', in particular; and it presumes an acquaintance with a particular region of space, with its contents, enabling the percipient to identify the particular cat and mat as belonging to the set of domestic furnishings to which implicit reference is being made (for presumably the statement is not just about any unspecified cat being on any unidentified mat).

In short, the achievement that constitutes an act of perception of this kind is one that presupposes an already, to some extent, devel-

3. See also A. M. Quinton, "The Problem of Perception," *Mind,* LXIV (1955), 28–51.

oped body of background knowledge, by reference to which what is presented to the senses is recognized and interpreted. Apart from this no perceptual achievement would be possible.

From a somewhat different angle a similar conclusion is reached by Professor R. J. Hirst in his book *The Problems of Perception*.[4] He considers perception to be a relation between a person and other public objects and events, which involves mental activity and so presents an inner and an outer aspect. On the inner side it is the '(awareness or consciousness of) an external object', which must be distinguished as 'a whole activity of the person'. To the person who is perceptually conscious the content of his consciousness is an external object or scene. In short, perception is essentially an intentional activity. Professor Hirst rejects any theory of sense data and all suggestion of private objects, though he admits perceptual errors and hallucinations. For him perception is a matter of degree, an activity which may be performed more or less successfully, in which the percipienda are elaborated in various ways and to different extents in order to produce the percepta. In describing this process, he follows the psychologists, who assert that the percipient achieves his percepta by the use of 'cues' and correlations which modify the purely sentient elements of his experience, largely in accordance with past learning. I have argued elsewhere [5] that the use of 'cues' and correlations (some psychologists even speak of 'assumptions' and 'hypotheses' in this connexion) cannot be reduced to mere physical or physiological processes, but is essentially epistemological implying an activity of reference and interpretation, which again implies a body of more or less developed and organized knowledge to which reference is made and in terms of which presentations are interpreted. It is not necessary, and would, in fact, be misguided, to regard this process as intellectual in the sense that it requires the explicit use of abstract concepts and ratiocination. The suggestion, attributed to Blanshard,[6] that it is a thought process, and the view advanced by Helmholtz and some of the older psychologists that it is 'unconscious inference' is commonly criticized (and insistently by

4. R. J. Hirst, *The Problems of Perception* (London: Allen and Unwin, 1959).

5. E. E. Harris, *The Foundations of Metaphysics in Science* (New York: Humanities Press, 1965), pp. 407 ff.

6. Brand Blanshard, *The Nature of Thought* (London: Allen and Unwin, 1939), vol. I.

Hirst) for this reason. But these writers, Blanshard especially (who should be given credit for drawing attention to this essentially mediate character of perception long before the others whom I have mentioned), by insisting that the process is 'implicit' or 'unconscious' themselves recognize its nonintellectual character. Unless we are to revert to sense-data (and perhaps even then) the recognition of some such process of mediation is inescapable. Equally inescapable is its logical and epistemological character as reference of a focal object to a context, articulated in some degree, spatio-temporally extended, and significant for the perceiving subject as a body of knowledge, understood in a dispositional sense.

Corroboration of this conclusion comes from another quarter when we turn to the work of Merleau-Ponty, for whom (as for Spinoza) perception is a functioning of the body as a whole, and not just the sense-organs. But the organism is sensitive not to separate, atomic, stimuli, but to a whole situation constituted by their constellation. Even below the level of self-awareness and intentionality proper, he points out, it displays behaviors "as if oriented towards the meaning of certain elementary situations." The significant word is "meaning." This orientation is no merely physical or physiological adjustment but involves an apprehension of the structure of the presented situation. Every perception, he says, takes place within a certain 'horizon' or setting (*entourage*) and ultimately in 'the world'; the perceived object itself being a unified and articulated texture of qualities and relations. If it is a visual spatial object it is given as "an infinite sum of an indefinite series of perspectival views," the interrelation of which is developed in a definite, recognizable order or form, so that the perceived thing is an inexhaustible system (*ensemble*). Consequently, perception is a synthesis performed by a subject able to delimit perspectival aspects and to transcend the immediately given—a reference to a whole which can be grasped only through certain of its parts.

In short, the percept is an object apprehended in a setting as belonging to a world which must therefore be in some way appreciated as a whole, or in general terms, however indefinite; and the logically prior condition of perceiving objects is the possession of some overall idea of the world, however vague, to which they belong and in which they are integrants. Observation, the fruit of perception, is thus embedded in and inseparable from interpretation, even at the level of common sense.

Science grows out of common sense and is continuous with it, in that the common sense level is simply the progressive development and clarification of the interpretation of percepts through the perceptual and intellectual exploration of the setting which gives them significance, and science carries this process to higher levels by fuller, more explicit and more precise exercise of the intellectual factor. *Pari passu* the observational aspect is sharpened and refined and becomes more accurate.

For scientific observation, moreover, the dependence on background knowledge is especially indispensable. The untutored savage presented with the instruments of a scientific laboratory in an experimental situation, can make no scientific observation whatsoever. Even the educated layman can understand what is happening with great difficulty and only with the help and explanation of the expert. Before any quantitative measurement can be made the experimental situation as a whole has to be understood: what is being measured? in what connexions? for what purposes? No instrument reading has significance unless its place in the experimental situation is known and its relevance to the questions that the experimenter is trying to answer. Only in answer to questions is any observation significant, and the questions arise only out of a theory already entertained and a body of scientific knowledge already achieved.

Moreover, what precisely is being observed in any experiment can be known only to one who is versed in the science concerned. That a particular pointer stands opposite a particular number or mark on a scale is altogether insignificant unless one knows that these things are part of, say, a galvanometer. That the indicated figure gives the measure of resistance to an electrical current cannot be appreciated unless the observer understands the principles on which galvanometers are constructed and to what this one in particular is connected. Clearly, a considerable body of knowledge of physical theory is presupposed in any one reading made with an instrument of this kind.

This is true at any level of scientific development. Observation is observation only if it answers relevant questions, these questions arise only out of prior knowledge—they are the ἀπορίη of theories already adopted or at least entertained. Apart from them, even though the objects are present to be observed, the percipient will not notice them—i.e. he will *not* observe them; and if he does, what he observes will depend upon the extent to which it is enlightened and interpreted by means of the theories to which the observation is contributing material. In short, theory is always, and always must be, prior to observation, and without it science would have no relevant empirical basis.

In any given situation the observable facts are infinite in number. Even those which are relevant to a particular problem may be innumerable. The scientist cannot observe them all and does not observe at random. To do so is strictly impossible, for the reasons already given, for to achieve knowledge of the observable facts one has to bring to bear upon them knowledge already acquired. He must, therefore, select and discriminate, and this he can do only in the light of the theories and hypotheses he has in mind and only so far as he can make his observations significant in reference to them.

This is well known to scientists. L. Bolzmann the founder of modern thermodynamics complained that "the lack of clarity in the principles of mechanics seems to be connected with the fact that one did not at once start with hypothetical pictures framed by our minds, but tried to start from experience." [7] We have also the more recent evidence of Einstein and Eddington. The former writes:

> By and by I despaired of the possibility of discovering the true laws by means of constructive efforts based on known facts. The longer and the more despairingly I tried, the more I came to the conviction that only the discovery of a universal formal principle could lead us to assured results. . . . How then could such a universal principle be found? (There follows the description of a piece of imaginative thinking impossible to test in any experiment which gave him "the germ of the special theory of relativity.") [8]

Eddington is still more positive:

> A scientist commonly professes [he writes] to base his beliefs on observation, not theories. Theories, it is said, are useful in suggest-

7. I. Bolzmann, *Vorlesungen über die Principen der Mechanik* (Leipzig, 1897), p. 2.

8. P. A. Schilpp (ed.), *Albert Einstein, Philosopher Scientist* (New York: Harper, 1959), pp. 52–53.

ing new ideas and new lines of investigation for the experimenter; but "hard facts" are the only proper ground for conclusion. I have never come across anyone who carries this profession into practice —certainly not the hard-headed experimentalist, who is more swayed by his theories because he is less accustomed to scrutinize them. Observation is not sufficient. We do not believe our eyes unless we are first convinced that what they appear to tell us is credible. . . . For the reader resolved to eschew theory and admit only definite observational facts, *all* astronomical books are banned. *There are no purely observational facts about the heavenly bodies.*[9]

But lest we should think these physicists prejudiced by their preoccupation with mathematics (though it is clear that this is not what inspires their remarks) let us hear that great biological observer Charles Darwin: "Without hypotheses," he says, "there can be no useful observation." [10]

This is why the method of science is misconceived by those who persist in the view that both for knowledge in general and the natural sciences in particular, sense observation is primary and self-contained, and must somehow be used as an independent check upon the accuracy of theories; and that the hall-mark of scientific method in natural science is inductive reasoning from particular matters of fact to general laws, an invalid process of inference which, Karl Popper rightly asserts, is actually never used in science and cannot be validated in logic.

Scientific thinking, or for that matter thinking at any level worthy of the name, is the systematic development of a conceptual (or theoretic) scheme covering a field of investigation in terms of which percepta, or observations, are interpreted. Without such a scheme we cannot even begin, and at every level of scientific development an accepted scheme is already at hand. This does not mean that it has its source in some mysterious *a priority* independent of experience. It is itself experience as it has been organized and developed up to that point; and experience is primitively sentient, and sentience becomes percipience only through organization. To ask which is

9. Sir Arthur S. Eddington, *The Expanding Universe* (Cambridge: The University Press, 1933), p. 17.

10. Quoted by W. E. le Gros Clark in "Anatomical Perspectives in Neuropsychiatry," *Perspectives in Neurophysiology*, ed. D. Richter.

temporally prior in this development, sense or thought, is to ask which came first the hen or the egg.

I shall not attempt to excavate origins but shall try to illustrate how, at a relatively elementary stage of scientific development, theory serves as the organizing factor in the construction of a coherent conception of the subject matter. The night sky presents the appearance of a myriad of bright points in a dark background, along with the moon as a superior luminary. The immediate presentation is one of completely random arrangement but more careful attention reveals recognizable patterns of spatial relation, some of which, if the observation is prolonged remain constant while others vary, so that over a protracted period the stars can be discovered to move across the heavens in various ways, some in circles, some along arcs which are greater or smaller segments of a circle, some in looping, oscillating courses, some faster and some more slowly. At first sight this is a bewildering variety; but these apparent movements (which, let us note, do not appear to the senses, but only to the thinking mind ordering the sensuous appearances), if they are related to the position of the sun at its rising and setting, and if we hypothesize that the heaven is a great globe, at the centre of which is the earth in the form of a relatively small sphere, and that the greater celestial sphere rotates about the small one on an axis, one end of which is marked by the pole-star, we can in large measure reduce this variety to order and regularity; and by marking out among the stars the path of the sun around the celestial sphere during the twelve months of the year (namely, the ecliptic), we get a coherent conception which will account for the movement of the majority of the heavenly bodies. This conception (sometimes called the two-sphere view) was, roughly, the earliest astronomical theory, and the history of astronomy, since it was adopted, up to the beginning of the 19th century was, almost exclusively, that of successive attempts to develop it so as to account for those aberrant motions which this simple theory leaves inexplicable, as contradictions within it—e.g. the planetary movements and the intermittent appearance of comets.

Each science is a similar sustained effort to comprehend an area or aspect of human experience in terms of a conceptual system which will reduce the relative chaos of percipient experience to coherent order. Success in this undertaking constitutes explanation. A fact or event is explained when it is seen in systematic context and is interpreted in terms of a systematic set of concepts. Explanation is not just 'analysis', in the sense of 'taking to pieces' or resolving into elements. Such analysis may be involved in explanation but by itself it explains nothing, unless the structural principles which originally held the elements together, into which the analysandum has been resolved, can also be demonstrated. Explanation is always synthesis just as much as analysis.

As the sciences progress they react upon one another and the effort becomes general to relate all their conceptual schemes and coordinate them into a single all-embracing theory. To describe the matter thus, however, is misleading, for in the nature of the case and as a matter of historical fact, the overarching conception is there from the beginning, first as a halting, naive and largely pictorial synthesis, but becoming more conceptual, more systematic and better articulated as the sciences develop. The discipline that produces this finally comprehensive synoptic conception is metaphysics which began in the west with Thales and Anaximander, gave birth to the sciences one by one, stimulated their development and profited by it, so that each owed its successive advances to the progress of the other, and the historical association, which I noted earlier, was natural and necessary.

Accordingly metaphysical explanation is the same in character as, and is only an extension of, scientific explanation, and metaphysical method is in principle a development and continuation of scientific method, as the latter may be said to be a specification, in different fields, of the former. The aim of metaphysics is a comprehensive conceptual scheme in the light of which the whole of experience can be organized and become intelligible. It is for this reason sometimes said (e.g. by Whitehead) to be concerned with the most general and pervasive features of the world and at others (e.g. by Plato) to be the most synoptic of the sciences. But, to quote A. E. Taylor again, "Metaphysics does not profess to deal with a certain group of facts lying outside the province of the 'sciences', but to deal with the same facts which form that province from a point of view which is not that of the experimental sciences." [11] The phrase "the sciences"

11. *Op. cit.*, p. 10.

is meant here to refer to all sciences, not just "the natural sciences" if that means only physics, chemistry and biology. Not only must the social sciences and the sciences of the mind be included, but whatever others may be necessary to deal with any field or aspect of experience; for, as has been said, metaphysics seeks a conceptual scheme comprehensive of the whole of experience.

The differences between scientific and metaphysical method derive only from this difference of scope. The self-limitation of the special sciences permits, and is defined by, initial assumptions, which the lack of such limitation in metaphysics forbids. The sciences do not examine or call these assumptions in question, but metaphysics must do so and may not itself adopt any, except provisionally. So Plato maintained that Dialectic eliminated hypotheses and was ἀνυπόθετος, and Kant declared any hypothesis in metaphysics contraband. Moreover, hypotheses are of two kinds, both in measure relative to the limitation in scope of a science. There are working hypotheses, which scientists adopt consciously and deliberately, and which direct and canalize their researches; and there are more fundamental presuppositions, often made as a matter of custom or tradition and without explicit reflection: such, for instance, as that every event has a cause or that all causation is efficient. For metaphysics all such hypotheses are contraband and the critical examination of them is consequently part of the metaphysician's business.

Similarly, in the special sciences, theory is more directly related to observation, whereas metaphysical theory is related to the empirical evidence indirectly through the special sciences. This, to the casual observer, may give the impression that metaphysics is purely speculative and "deductive" in its method "scorning the base degrees by which it did ascend" and producing empirically unverifiable theses. But this impression is false. All science is speculative. If I may quote a scientist once more, Professor le Gros Clark says: "Contrary to views which are sometimes expressed, speculation is by no means to be eschewed in scientific pursuits—indeed it is essential for progress since it is the generator of ideas and ideas stimulate enquiry." [12] In this respect Metaphysics does not differ from science, nor does it differ in its manner of verification.

12. le Gros Clark, *loc. cit.*

Theory and observation in science are not two separate and independent factors by one of which we may check or test the other. A scientific theory is a more or less organized and coherent interpretation of what is observed, without which the observed facts lose their character and significance. Observation, we have already noted, is saturated with interpretation, of which theory is no more than the development. Theory and observation together form a single whole of knowledge. But this is only 'more or less' coherently organized and the endeavour of science is constantly to improve the system so that it becomes more comprehensive and self-consistent. Verification, therefore, always consists in the assembling of mutually corroborative evidence, the interconnections of which make the denial of the theory impossible without the breakdown of the entire conceptual scheme. Falsification consists in the failure of corroboration, which demands either modification of the theory or the discovery of new connexions which will reconcile the conflicting evidence. The process, in science, obviously involves observation for without it theory would have no subject matter. In the case of astronomy, if no stars were observed there would be no science, but the two-sphere view cannot be 'verified' by comparison with some independently observed body of facts, it is simply the way in which the observed facts are organized. The Copernican system is another way of organizing them which can explain more satisfactorily planetary and cometary movements, as the earlier theories could not. Verification in science is, therefore, by mutual corroboration of evidence.[13]

Metaphysics is no more and no less than an attempt to organize comprehensively the deliverances of the sciences into a single world-view by means of a universal conceptual scheme in which all sciences can be integrated. That this can be adequately or successfully done without clarification and analysis of concepts is hardly to be ex-

13. Karl Popper has implicitly subscribed to a similar view in a significant footnote in *The Logic of Scientific Discovery* (New York: Basic Books, 1959), p. 87, "If I assert that there is a family of white ravens in the New York zoo, then I assert something which can be tested *in principle*. If somebody wishes to test it, and is informed, upon arrival, that the family has died, or has never been heard of, it is left to him to accept or reject my falsifying . . . statement. As a rule, he will have means of forming an opinion by examining witnesses, documents, etc.; that is to say by appealing to other intersubjectively testable and reproducible facts."

pected; so part of the metaphysician's task, and part of the explanatory function of metaphysical theory, will be conceptual analysis, which will naturally consider the ways in which terms and concepts are used, and how they function in science. But metaphysics cannot be restricted to such analysis, which is ancillary to its constructive purpose of providing a coherent explanatory scheme in which all possible sciences can themselves find place.

Moreover, the extension of scope involved in metaphysical explanation is bound to require the modification of the use and application of concepts, and what is appropriate in one science may well prove unacceptable in another and still more in a wider context. The term 'evolution,' for instance, and its connotation are not the same in physics (e.g. in "the evolution of the galaxy") as in biology; and, as Collingwood pointed out, when the idea is given philosophical significance it has to be modified still further.[14] Analysis of concepts, therefore, will also involve criticism, and adjustment in the light of such criticism. Consequently, metaphysics is inevitably "revisionary"; and because the use of a concept in each science is not separate and independent of its use in the others, and because the transition from one science to the next can be represented as a continuous development, this revisionary function is essential to the descriptive, which cannot be adequately or accurately carried out without it.

Further, this process of analysis and criticism of concepts and presuppositions is bound to call in question the credentials and assess the claims of the sciences concerned. Metaphysics, therefore like all philosophy will be what Collingwood called a criteriological science and will have a normative as well as a descriptive character. Moreover, as it seeks what we may now, without mystification, call ultimate explanation, it will be concerned with and will inquire into the ultimate grounds, whatever they be and if there are any (questions metaphysicians must raise), of the existence and nature of the subject matter of all the special sciences that fall under its purview.

The method of verifying a metaphysical theory, therefore, is to seek corroborative evidence among the sciences. It will, naturally, be for the most part theoretical evidence, itself an interpretation of

---

14. R. J. Collingwood, *An Essay on Philosophical Method* (London: Oxford University Press, 1933), p. 34.

observed fact; but the verificatory process is no different in principle from that of science. To neglect the evidence, as science provides it would be fatal, and, through the mediation of science, this evidence is as 'empirical' as any need be. Metaphysics, then, is an interpretation of interpretations; it is science of the second, or possibly of an even higher, degree. It is, in short, meta-science—and what else did the word "meta-physics" ever legitimately mean?

Albert Einstein

Albert Einstein (1879–1955), creator of the special and general theories of relativity, a leader in the early development of quantum theory and atomic physics, and the best known scientist of contemporary times, is popularly linked with Isaac Newton as the two pivotal figures in the development of physics. The parallels are striking. Both men were lonely thinkers, preferring to work in isolation on the issues that intrigued them. Both men did their most creative work at the age of twenty-six. Einstein's views on space, time, matter, and energy replaced Newton's. Yet the excerpts selected from each man's writings show a strikingly dissimilar view of the nature of scientific explanation. Newton stressed observation and induction and regarded speculative hypotheses with suspicion. Einstein, especially in his later years, came more and more to stress the free creative role of the scientist in fashioning concepts and introducing speculative hypotheses. That such free creations of the human mind could correspond to, and in some sense capture, physical reality was, in Einstein's view, something of a miracle; but on this miracle rested the possibility of scientific explanation. The brief selection reproduced here represents his contribution to a symposium on the philosophy of Bertrand Russell. He used this as an occasion for presenting one of the briefest and clearest expositions of his own developed views on the nature of scientific explanation, its correspondence to reality, and its relation to metaphysics.

Remarks on Bertrand Russell's
Theory of Knowledge

When the editor asked me to write something about Bertrand Russell, my admiration and respect for that author at once induced me to say yes. I owe innumerable happy hours to the reading of Russell's works, something which I cannot say of any other contemporary scientific writer, with the exception of Thorstein Veblen. Soon, however, I discovered that it is easier to give such a promise than to fulfill it. I had promised to say something about Russell as philosopher and epistemologist. After having in full confidence begun with it, I quickly recognized what a slippery field I had ventured upon, having, due to lack of experience, until now cautiously limited myself to the field of physics. The present difficulties of his science force the physicist to come to grips with philosophical problems to a greater degree than was the case with earlier generations. Although I shall not speak here of those difficulties, it was my concern with them, more than anything else, which led me to the position outlined in this essay.

In the evolution of philosophic thought through the centuries the following question has played a major role: What knowledge is pure thought able to supply independently of sense perception? Is there any such knowledge? If not, what precisely is the relation between our knowledge and the raw-material furnished by sense-impressions? An almost boundless chaos of philosophical opinions corresponds to these questions and to a few others intimately connected with them. Nevertheless there is visible in this process of relatively fruitless but heroic endeavours a systematic trend of

From *The Philosophy of Bertrand Russell*, pp. 279–91. Now published by The Open Court Publishing Company, La Salle, Illinois. By permission of the publisher. Translated from the original German by Paul Arthur Schilpp.

development, namely an increasing scepticism concerning every attempt by means of pure thought to learn something about the 'objective world', about the world of 'things' in contrast to the world of mere 'concepts and ideas'. Be it said parenthetically that, just as on the part of a real philosopher, quotation-marks are used here to introduce an illegitimate concept, which the reader is asked to permit for the moment, although the concept is suspect in the eyes of the philosophical police.

During philosophy's childhood it was rather generally believed that it is possible to find everything which can be known by means of mere reflection. It was an illusion which anyone can easily understand if, for a moment, he dismisses what he has learned from later philosophy and from natural science; he will not be surprised to find that Plato ascribed a higher reality to 'ideas' than to empirically experienceable things. Even in Spinoza and as late as in Hegel this prejudice was the vitalizing force which seems still to have played the major role. Someone, indeed, might even raise the question whether, without something of this illusion, anything really great can be achieved in the realm of philosophic thought—but we do not wish to ask this question.

This more aristocratic illusion concerning the unlimited penetrative power of thought has as its counterpart the more plebeian illusion of naïve realism, according to which things 'are' as they are perceived by us through our senses. This illusion dominates the daily life of men and of animals; it is also the point of departure in all of the sciences, especially of the natural sciences.

The effort to overcome these two illusions is not independent the one of the other. The overcoming of naïve realism has been relatively simple. In his introduction to his volume, *An Inquiry Into Meaning and Truth,* Russell has characterized this process in a marvellously pregnant fashion:

> We all start from "naïve realism," i.e., the doctrine that things are what they seem. We think that grass is green, that stones are hard, and that snow is cold. But physics assures us that the greenness of grass, the hardness of stones, and the coldness of snow, are not the greenness, hardness, and coldness that we know in our own experience, but something very different. The observer, when he seems to himself to be observing a stone, is really, if physics is to be believed, observing the effects of the stone upon himself. Thus

science seems to be at war with itself: when it most means to be
objective, it finds itself plunged into subjectivity against its will.
Naïve realism leads to physics, and physics, if true, shows that
naïve realism is false. Therefore naïve realism, if true, is false;
therefore it is false. (pp. 14–15).

Apart from their masterful formulation these lines say something
which had never previously occurred to me. For, superficially considered, the mode of thought in Berkeley and Hume seems to
stand in contrast to the mode of thought in the natural sciences.
However, Russell's just cited remark uncovers a connection: If
Berkeley relies upon the fact that we do not directly grasp the
'things' of the external world through our senses, but that only
events causally connected with the presence of 'things' reach our
sense-organs, then this is a consideration which gets its persuasive
character from our confidence in the physical mode of thought.
For, if one doubts the physical mode of thought in even its most
general features, there is no necessity to interpolate between the
object and the act of vision anything which separates the object
from the subject and makes the 'existence of the object' problemati-
cal.

It was, however, the very same physical mode of thought and
its practical successes which have shaken the confidence in the possibility of understanding things and their relations by means of
purely speculative thought. Gradually the conviction gained recognition that all knowledge about things is exclusively a working-
over of the raw-material furnished by the senses. In this general
(and intentionally somewhat vaguely stated) form this sentence is
probably today commonly accepted. But this conviction does not
rest on the supposition that anyone has actually proved the impossibility of gaining knowledge of reality by means of pure speculation, but rather upon the fact that the empirical (in the above
mentioned sense) procedure alone has shown its capacity to be the
source of knowledge. Galileo and Hume first upheld this principle
with full clarity and decisiveness.

Hume saw that concepts which we must regard as essential,
such as, for example, causal connection, cannot be gained from
material given to us by the senses. This insight led him to a sceptical
attitude as concerns knowledge of any kind. If one reads Hume's

books, one is amazed that many and sometimes even highly esteemed philosophers after him have been able to write so much obscure stuff and even find grateful readers for it. Hume has permanently influenced the development of the best of philosophers who came after him. One senses him in the reading of Russell's philosophical analyses, whose acumen and simplicity of expression have often reminded me of Hume.

Man has an intense desire for assured knowledge. That is why Hume's clear message seemed crushing: The sensory raw-material, the only source of our knowledge, through habit may lead us to belief and expectation but not to the knowledge and still less to the understanding of law-abiding relations. Then Kant took the stage with an idea which, though certainly untenable in the form in which he put it, signified a step towards the solution of Hume's dilemma: Whatever in knowledge is of empirical origin is never certain (Hume). If, therefore, we have definitely assured knowledge, it must be grounded in reason itself. This is held to be the case, for example, in the propositions of geometry and in the principle of causality. These and certain other types of knowledge are, so to speak, a part of the instrumentality of thinking and therefore do not previously have to be gained from sense data (i.e., they are *a priori* knowledge). Today everyone knows of course that the mentioned concepts contain nothing of the certainty, of the inherent necessity, which Kant had attributed to them. The following, however, appears to me to be correct in Kant's statement of the problem: in thinking we use, with a certain 'right', concepts to which there is no access from the materials of sensory experience, if the situation is viewed from the logical point of view.

As a matter of fact, I am convinced that even much more is to be asserted: the concepts which arise in our thought and in our linguistic expressions are all—when viewed logically—the free creations of thought which cannot inductively be gained from sense-experiences. This is not so easily noticed only because we have the habit of combining certain concepts and conceptual relations (propositions) so definitely with certain sense-experiences that we do not become conscious of the gulf—logically unbridgeable—which separates the world of sensory experiences from the world of concepts and propositions.

Thus, for example, the series of integers is obviously an invention of the human mind, a self-created tool which simplifies the ordering of certain sensory experiences. But there is no way in which this concept could be made to grow, as it were, directly out of sense experiences. It is deliberately that I choose here the concept of numbers, because it belongs to pre-scientific thinking and because, in spite of that fact, its constructive character is still easily recognizable. The more, however, we turn to the most primitive concepts of everyday life, the more difficult it becomes amidst the mass of inveterate habits to recognize the concept as an independent creation of thinking. It was thus that the fateful conception—fateful, that is to say, for an understanding of the here existing conditions—could arise, according to which the concepts originate from experience by way of 'abstraction', i.e., through omission of a part of its content. I want to indicate now why this conception appears to me to be so fateful.

As soon as one is at home in Hume's critique one is easily led to believe that all those concepts and propositions which cannot be deduced from the sensory raw-material are, on account of their 'metaphysical' character, to be removed from thinking. For all thought acquires material content only through its relationship with that sensory material. This latter proposition I take to be entirely true; but I hold the prescription for thinking which is grounded on this proposition to be false. For this claim—if only carried through consistently—absolutely excludes thinking of any kind as 'metaphysical'.

In order that thinking might not degenerate into 'metaphysics', or into empty talk, it is only necessary that enough propositions of the conceptual system be firmly enough connected with sensory experiences and that the conceptual system, in view of its task of ordering and surveying sense-experience, should show as much unity and parsimony as possible. Beyond that, however, the 'system' is (as regards logic) a free play with symbols according to (logical) arbitrarily given rules of the game. All this applies as much (and in the same manner) to the thinking in daily life as to the more consciously and systematically constructed thought in the sciences.

It will now be clear what is meant if I make the following statement: By his clear critique Hume did not only advance philosophy in a decisive way but also—though through no fault of his—created a danger for philosophy in that, following his critique, a fateful 'fear of metaphysics' arose which has come to be a malady of contemporary empiricistic philosophizing; this malady is the counterpart to that earlier philosophizing in the clouds, which thought it could neglect and dispense with what was given by the senses.

No matter how much one may admire the acute analysis which Russell has given us in his latest book on *Meaning and Truth,* it still seems to me that even there the spectre of the metaphysical fear has caused some damage. For this fear seems to me, for example, to be the cause for conceiving of the 'thing' as a 'bundle of qualities', such that the 'qualities' are to be taken from the sensory raw-material. Now the fact that two things are said to be one and the same thing, if they coincide in all qualities, forces one to consider the geometrical relations between things as belonging to their qualities. (Otherwise one is forced to look upon the Eiffel Tower in Paris and that in New York as 'the same thing'.) Over against that I see no 'metaphysical' danger in taking the thing (the object in the sense of physics) as an independent concept into the system together with the proper spatio-temporal structure.

In view of these endeavours I am particularly pleased to note that, in the last chapter of the book, it finally crops out that one can, after all, not get along without 'metaphysics'. The only thing to which I take exception there is the bad intellectual conscience which shines through between the lines.

Wilfrid Sellars

Wilfrid Sellars inherited an interest in the problem of scientific realism from his father, Roy Wood Sellars, who joined Lovejoy and Santayana as one of the leaders in the American school of critical realism. Although Wilfrid Sellars's position on scientific realism is not essentially different from his father's, his development of the position is different. Thanks to his studies at Oxford and his interest in logical positivism, Sellars mastered the techniques of linguistic and logical analysis and uses these techniques to develop a doctrine of critical realism with a unique subtlety and epistemological sophistication. While this often makes for rather difficult reading, it serves to set critical standards for informed discussion. In recent years Sellars's writings have manifested a growing concern with philosophical problems of perennial interest in metaphysics, ethics, rational psychology, and the history of philosophy. He is presently a professor of philosophy at the University of Pittsburgh. The following essay represents one of his most important writings on the nature of scientific explanation and the problem of scientific realism.

The Language of Theories [1]

I. Introduction

1. My purpose is to see what fresh light, if any, is thrown on old familiar puzzles about the empirical (or factual) meaningfulness of theoretical statements and the reality of theoretical entities by certain views on related topics which I have sketched in Chapters 5, 8, and 11 and in a recent paper.[2] These views concern (a) the interpretation of basic semantical categories; (b) the role of theories in scientific explanation.

2. The term 'theory', it is generally recognized, covers a wide variety of explanatory frameworks resembling one another by that family resemblance which is easy to discern but most difficult to describe. Each type of theory presents its own problems to the philosopher of science, and although current literature shows an increasing tendency to reflect the realities of scientific practice rather than antecedent epistemological commitments, the type of theory with which I shall be concerned—namely, that which postulates unobserved entities to explain observable phenomena—is still

From CURRENT ISSUES IN THE PHILOSOPHY OF SCIENCE edited by Herbert Feigl and Grover Maxwell. Copyright © 1961 by Holt, Rinehart and Winston, Inc. Reprinted by permission of Holt, Rinehart and Winston, Inc.

1. I wish to acknowledge the invaluable assistance I have received from friends and colleagues who gave me their comments on an earlier draft of this chapter. I am particularly grateful to Professor Adolf Grünbaum for a page by page critique with respect to both exposition and substance, without which this chapter would have fallen far shorter than it does of saying what I wanted it to say.

2. "Counterfactuals, Dispositions and the Causal Modalities," in Volume II of *Minnesota Studies in the Philosophy of Science*, edited by H. Feigl, M. Scriven, and G. Maxwell, and published by the University of Minnesota Press (Minneapolis: 1958).

suffering from the effects of a Procrustean treatment by positivistically oriented philosophies of science.

3. I shall assume, at least to begin with, that *something* like the standard modern account of this type of theory is correct. And in view of the distinguished names associated with this account, it would be most surprising if it were not close to the truth. It is built upon a distinction between: (*a*) the vocabulary, postulates, and theorems of the theory as an uninterpreted calculus; (*b*) the vocabulary and inductively testable statements of the observation framework; (*c*) the 'correspondence rules' which correlate, in a way which shows certain analogies to inference and certain analogies to translations, statements in the theoretical vocabulary with statements in the language of observation. Each of these categories calls for a brief initial comment.

4. The theoretical language contains, in addition to that part of vocabulary which ostensibly refers to unobserved entities and their properties, (*a*) logical and mathematical expressions which have their ordinary sense, and (*b*) the vocabulary of space and time. (Query: can we say that the latter part of the theoretical vocabulary, too, has its ordinary sense? To use the material mode, are the space and time of kinetic theory the same as the space and time of the observable world, or do they merely 'correspond' to them? In relativity physics it is surely the latter.)

5. The nontheoretical language with which a given theory is connected by means of correspondence rules may itself be a theory with respect to some other framework, in which case it is nontheoretical only in a relative sense. This calls up a picture of levels of theory and suggests that there is a level which can be called nontheoretical in an absolute sense. Let us assume for the moment that there is such a level and that it is the level of the observable things and properties of the everyday world and of the constructs which can be explicitly defined in terms of them. If following Carnap we call the language appropriate to this level the *physical-thing language,* then the above assumption can be formulated as the thesis that the physical-thing language is a nontheoretical language in an absolute sense. The task of theory is then construed to be that of explaining inductively testable generalizations formulated in the physical-thing language, which task is equated with *deriving* the latter from the theory by means of the correspondence rules.

6. Correspondence rules typically connect defined expressions in the theoretical language with definable expressions in the language of observation. They are often said to give a 'partial interpretation' of the theory in terms of observables, but this is at best a very misleading way of talking; for whatever may be true of 'correspondence rules' in the case of physical geometry,[3] it is simply not true, in the case of theories which postulate unobserved micro-entities, that a correspondence rule stipulates that a theoretical expression is to *have the same sense* as the correlated expression in the observation language. The phrase 'partial interpretation' suggests that the only sense in which the interpretation fails to be a translation is that it is partial; that is, that while *some* stipulations of identity of sense are laid down, they do not suffice to make possible a complete translation of the theoretical language into the language of observation. It is less misleading to say that while the correspondence rules *co-ordinate* theoretical and observational sentences, neither they nor the derivative rules which are their consequences place the primitives of the theory into one-man correspondence with observation language counterparts. This way of putting it does not suggest, as does talk of 'partial interpretation', that if the partial correlation could be made complete, it would *be* a translation. (That in some cases a 'complete correlation' might be *transformed* into a translation by reformulating correspondence rules as semantical stipulations is beside the point.)

7. For the time being, then, we shall regard the correspondence rules of theories of the kind we are examining as a special kind of verbal bridge taking one from statements in the theoretical vocabulary to statements in the observation vocabulary and vice versa. The term 'correspondence rule' has the advantage, as compared with 'bridge law', 'co-ordinating definition', or 'interpretation', of being neutral as between various interpretations of the exact role played by these bridges in different kinds of theory.

8. Puzzles about the meaning of theoretical terms and the reality of theoretical entities are so intimately bound up with the status of correspondence rules that to clarify the latter would almost auto-

---

3. The case of geometry is not independent, for geometrical concepts must be defined for micro-entities. Even if abstraction is made from this, there remains the problems of extending idealized congruences to situations in which the congruences are physically impossible—for example, the centre of the sun.

matically resolve the former. This fact is the key to my strategy in this chapter. But before attempting to develop a suitable framework for this analytical task, a few remarks on contemporary treatments of correspondence rules are in order. Until recently it was customary, in schematic representations of theories, to keep the postulates and theorems of the theory, the empirical generalizations of the observation framework, and the correspondence rules linking theory with observation in three distinct compartments. This had the value of emphasizing the methodologically distinct roles of these three different types of statement. On the other hand this mode of representation carried with it the suggestion of an *ontological* (as contrasted with methodological) dualism of theoretical and observational universes of discourse which a more neutral presentation might obviate. Thus it has recently been the tendency to list the correspondence rules with the postulates of the theory, distinguishing them simply as those postulates which include observational as well as theoretical expressions.[4] This procedure can do no harm if the relevant methodological and semantical distinctions ultimately find adequate expression in some other way.

II. Some Semantical Distinctions[5]

9. That 'meaning' has many meanings is an axiom of contemporary philosophy. Of these, some are logical in a narrow sense of the term—thus, naming, denoting, connoting. Others are logical in a somewhat more inclusive sense. Thus there is the methodological sense in which a meaningful expression may be scientifically meaningless, have no scientific point—thus an arbitrarily defined function of measurable quantities. Still other varieties of meaning, though

4. This method of presentation is in certain respects analogous to that of drawing no formal distinction between definitions, on the one hand, and postulates and theorems of the form "a = b" on the other in the development of a calculus, leaving it to subsequent reflection to determine how the latter are to be parcelled out into definitional and nondefinitional identities.

5. The substantive argument of the chapter resumes with Section III. The present section draws semantical distinctions which are later used to give a precise formulation to the problem of the reality of theoretical entities and its correct solution. It may, however, be omitted without prejudice to the main thrust of the argument, and should, perhaps, in any case be omitted on a first reading.

'semantical' in a broad sense of the term, are the concern of psychology and rhetoric rather than of logic in even the most inclusive sense of this term.

10. It is with those senses of meaning which directly pertain to logical theory in a narrow sense that I shall be concerned in this part of my argument. I shall attempt to sketch a coherent treatment of basic semantical categories which may throw light on questions of meaning and existence pertaining to theoretical discourse. I shall make no attempt to provide a formalized theory of meaning elegantly reduced to a minimum of primitive notions and propositions. Such attempts are premature and dangerous in any area if they are based on misinterpretations of the initial explicanda. Nowhere, in my opinion, have these dangers been realized more disastrously than in some recent theories of meaning.[6]

11. Thus, instead of attempting to explicate the various logical senses of 'meaning' in terms of a single primitive notion—for example, *the designation relation* or *denotes*—I shall simply give what I hope is a defensible account of the various kinds or modes of meaning and of how they are inter-related. I shall distinguish the following:

- (*a*)   Meaning as translation
- (*b*)   Meaning as sense
- (*c*)   Meaning as naming
- (*d*)   Meaning as connotation
- (*e*)   Meaning as denotation

12. The expression 'means' as a translation rubric is easily confused with its other uses. The essential feature of this use is that whether the translation be from one language to another, or from one expression to another in the same language, the translated

---

6. I have in mind (*Introduction to Semantics*) Carnap's formalization of semantical theory in terms of a primitive relation of designation which holds between words and *extralinguistic* entities. This reconstruction commits one to the idea that if a language is meaningful, there exists a domain of entities (the *designata* of its names and predicates) which exist independently of any human concept formation. Of course, Carnap's semantical theory as such involves no commitment as to what this domain includes, but if one adds the premise that the physical thing *language* is meaningful, one is committed to the idea that the framework of observable physical things and their properties has an absolute reality which, if the argument of the present paper is sound, it does not have.

expression and the expression into which it translates must have the same use.[7] Thus if we are to use 'means' in this sense we must say *not*

(1) (The English adjective) "round" means *circularity*

but

(2) (The English adjective) "round" means *circular*

where "round" and "circular" are both *predicative* expressions having the same sense. The essential difference between (2) and

(3) (The English adjective) "round" has the same use as (the English adjective) "circular"

is that to say of two expressions in a language that they have the same use—as in (3)—is not to *give* that use; (2) gives the use of "round" by presupposing that "circular" is in the active vocabulary of the person to whom the statement is made and, normally, of the person making the statements, whereas (3) does not.

13. The translation use of 'means' gives expression to the fact that the same linguistic role can be played by different expressions. It should perhaps be added that since statements involving this sense of 'means' are used to *explain* the use of one expression in terms of the use of others, they are usually not reversible. Thus,

(4) "triangle" means *plane figure bounded by three straight lines*

and

(5) "plane figure, etc." means *triangle*

are not equally appropriate. With this qualification, to say that an expression has meaning in this sense is to commit oneself to explaining its use by means of another expression with the same use. Such a statement as

(6) "Red" means *red*

---

7. To speak of two expressions as having the same use is to presuppose a criterion of sameness of use which separates relevant from irrelevant differences in use. Clearly, differences which are irrelevant to one context of inquiry may be relevant to another.

can perhaps be construed as a limiting case which gets its sense by implying that "red" *has* a use, and by implying that this use is not capable of explanation in terms of more basic expressions.

14. Closely related to 'means' as a translation rubric is 'means' in the sense of 'expresses the concept . . .' In this case we must say not

(2)   (The English adjective) "round" means *circular*

but

(1)   (The English adjective) "round" means *circularity*

or, as I shall put it,

(7)   (The English adjective) "round" *expresses the concept* circularity.[8]

Notice that it would be incorrect to put this by saying that

(8)   "Round" *names* the concept Circularity

for this is done by "roundness." Thus we have

(9)   "Round" means [9] *circular.*

(10)   "Round" expresses the concept Circularity
(11)   "Roundness" names the concept Circularity.

15. Again,

(12)   (The Italian word) "Parigi" *means* Paris

in the translation sense of 'means'. But here, of course, it is also true that

(13)   (The Italian word) "Parigi" *names* Paris

It is even true that

(14)   (The Italian word) "Parigi" *expresses the concept* Paris or, to use a medieval locution, Pariseity.

8. In what follows, I shall choose my examples of meaning statements of the first two kinds without regard to whether they would be good explanations of usage in standard conditions. I shall also drop the explicit reference to the language to which an expression belongs when the context makes it clear which langauge is intended.

9. From now on I shall use 'means' in examples only in the sense of the translation rubric.

For, as we shall see, we must recognize individual concepts as well as universal concepts (and, indeed, other kinds of concept as well), and can do so without ontological discomfort.

16. Again, we shall say that

(15)   "Parigi" *connotes* the property of being the capital of France and, in general, names connote those properties the possession of which are criteria for being properly referred to by the name in question. The distinction between the concept expressed by a name and the concepts or properties connoted by the name is of the utmost importance as illustrations will shortly make evident. The concept can, within limits, remain the same although the criteria change or become differently weighted.[10] Let me give two further sets of illustrations for the distinctions I have been drawing.

(16)   (The Italian word) "Icaro" means *Icarus*
(17)   (The Italian word) "Icaro" *names* nothing real (or actual)
(18)   (The Italian word) "Icaro" expresses the concept Icarus (or Icaruseity)
(19)   (The Italian word) "Icaro" connotes the property of being the son of Daedalus.

An illustration in terms of a common rather than a proper name would be

(20)   "Cheval" means *horse*
(21)   "Cheval" expresses the concept Horsekind
(22)   "Cheval" connotes the property of having four legs
(23)   "Cheval" names Man o' War, Zev, etc.

17. Finally, denoting must be distinguished from naming. Thus, we can say that

(24)   "round" *denotes* circular things

but does not name them. The closest thing to a name of circular things *qua* circular would be the common noun expression "circular

10. See Wittgenstein's discussion of essentially the same point in *Philosophical Investigations*, § 79. See also § 47 of my essay "Counterfactuals, Dispositions and the Causal Modalities."

thing." Again, for "round" to denote, in this sense, circular things is not the same as for it to denote the class of circular things. If we wish to use the term 'denotes' in such a way that

(25)   "round" denotes the class of circular things,

we must be careful to distinguish between *naming* a class and *denoting* a class, for "circular" is not the name of anything, let alone a class.

18. I shall return to the task of distinguishing and relating the above modes of meaning shortly. But first I want to point out that nothing would seem to be less controversial nor more trivial than that theoretical terms can be said to have meaning or be meaningful where the sense of 'meaning' envisaged is that which pertains to translation. Surely in the sense in which

(9)   "Round" means *circular*

so that there is something which "round" means—that is, *circular*—it is equally true that

(26)   (The German word) "Molekuel" means *molecule,*

so that there is something which "Molekuel" means—that is, *molecule*. All that would seem to be required is that such statements as (26) be made to people who have the theoretical expression which serves as *explicans* in their active vocabulary.

19. Notice, however, that he must have it in his active vocabulary as an expression which belongs to a theoretical language. Thus if theoretical expressions functioned *merely* as expressions in a *purely* syntactical game, which they obviously do not, it would be patently incorrect to bring them into *meaning* statements, whether as *explicandum* or as *explicans*.

20. But to make this point, sound though it be, is not to prove that theoretical terms in established theories 'have meaning' in any more interesting sense than that they are translatable. Nor, strictly, does it establish that they have meaning in even this limited sense. For we are immediately faced by the new question: When is an expression in what is prima facie a language functioning enough unlike a *mere* counter to warrant saying that we have *translated* it, as opposed to merely correlating it with another counter which we know how to deploy? For even if we were to decide that theoretical expressions were too much like pieces in a

game to be properly talked about in terms of meaning, we could explain their use by means of a rubric which might also be used in explaining an oddly shaped chess piece by correlating it with this piece in our set. We would then say that the use of translation talk in connection with theoretical expressions is best regarded as a metaphorical extension of the translation rubric to contexts which resemble translations in certain respects, but are not translations proper.

21. Thus the fact that it would be odd to deny that expressions in a French formulation of kinetic theory translate into their English equivalents is by no means a conclusive reason for holding that the language of kinetic theory is a language in the full-blooded sense of the term. Might not theoretical terms have meaning in the translational mode, without having it in any of the other modes we have distinguished? After all, even such a properly linguistic expression as the French "helas!" translates into "alas!" but certainly names nothing, has no connotation, and expresses no concept.[11] And "oui" translates into "yes."

22. If we turn first to meaning as expressing a concept, we must face the question as to what exactly is conveyed by such a statement as

(10)   "Round" expresses the concept Circularity

or

(26)   "Socrates" expresses the individual concept Socrateity.

Without further ado, I shall propose a straightforward, if radical, thesis to the effect that the sense of (10) is but little different from that of the translation statement

(9)   "Round" means *circular.*

The difference consists essentially in the fact that whereas in (9) the adjective "circular" is used to *give* the role shared by "round" and "circular" and does not mention the role by means of a *name* (though it implicitly describes it as the role played in our language by "circular"),[12] in (10) we find a *name* for this role, a name which

11. It may, nevertheless, be said to express a sense. See below, § 23.
12. See 12 above.

is formed in a special way from a sign design which plays the named role in our language.[13]

23. But if this account is correct, then to say that

(27)   "Molekuel" expresses the concept of a molecule

and, in general, to make statements of the form

(28)   ". . . ." expresses the concept—

about theoretical expressions is simply another way of drawing upon the fact that theory-language roles, like observation-language roles, can be played by more than one set of sign designs. We must, however, be a bit more discriminating than in our discussion of the translation rubric, in view of the fact that not all properly linguistic roles are *conceptual*. Thus, while it makes sense to say that "round" expresses the concept [property] Roundness, that "and" expresses the concept [operation] of Conjunction and that *"notwendig"* (in German) expresses the concept [modality] Necessity, etc., we can scarcely say that "Hélas!" (in French) expresses the concept Alas, though we *can* say

(29)   "Hélas!" (in French) expresses the sense Alas

For this gives expression to the fact that "Hélas!" plays in French the role that is played by the English word "Alas!" and is equivalent to

(30)   "Hélas!" (in French) plays the role played by "Alas!" in our language.

Just where we are to draw the line between expressions which express concepts, and those which, though properly linguistic, do

13. See my "Quotation Marks, Sentences and Propositions," in Volume X of *Philosophy and Phenomenological Research*, 1950. (Added in proof) Strictly speaking "circularity" is not the name of the role played by "circular" in English, but it rather to be construed as a singular term formed from a meta-linguistic common noun for items which play this role (e.g. there are several "circular's" on this page), as "the pawn" is formed from an object language common noun for items which play a certain role in chess. "Circularity," then, is the name of a linguistic type in the sense in which "the pawn" is the name of a chess piece. Similar considerations apply to other abstract singular terms, e.g. "that Chicago is large." For an elaboration of this interpretation of abstract singular terms, see my "Abstract Entities," in the June 1963 issue of *The Review of Metaphysics*.

not, is a nice question which I shall make no systematic attempt to answer.[14]

24. The above remarks may reconcile us to the idea that theoretical expressions can correctly be said to express concepts. But do they *name* anything? Do they *denote* anything?

25. Names, we have seen, connote criteria and name the objects which satisfy these criteria. We have distinguished between two radically different kinds of object which we may illustrate, respectively, by Socrates and by Roundness. Roughly the distinction is between those objects which are concepts and those which are not.[15]

26. Nonconceptual objects can be roughly divided into *basic* and *derivative*. Derivative objects can be informally characterized as those which are referred to by noun expressions that can be eliminated by contextual definition. In this sense events are derivative objects in the physical-thing framework. Statements about the events in which physical things participate can be reduced to statements in which all the nonpredicative expressions refer to physical things.[16] In the framework of kinetic theory, as classically presented, the basic objects (granted that we *can* speak of theoretical objects) would be individual molecules.

27. To know that a name names something is to know that some objects satisfies the connotation of the name and by doing so satis-

14. If one begins by listing a variety of types of expression which can without too much discomfort be said to express concepts—noun expressions, predicative expressions, logical connectives, quantifiers—one is likely to conclude that to express a concept is to be relevant to inferences which can be drawn from statements in which the expression occurs; and to note that "All men are mortal, alas!" has no more inferential force (pragmatic implications aside) than "All men are mortal." It would be because "good," "bad," "right," "wrong,' etc., do play roles in practical reasoning that they could properly be said to express concepts. I think that this approach is sound, but to be carefully worked out it would require a precise account of the difference between such an obviously nonconceptual expression as a left-hand parenthesis, and logical operators, such as "and." Ordinary distinctions between 'categorematic' and 'syncategorematic' expressions lump together a distinction of kind with distinctions of degree.

15. (Added in proof) My use of 'concept' corresponds closely to Frege's use of 'sense'. It is the predicative subset of concepts, as I am using this term, which correspond (differences of theory aside) for Frege's concepts. My 'concepts' are distributive objects in the sense in which the pawn is a distributive object. See footnote 6 on p. 187 above and the articles to which it refers.

16. See my essay "Time" in Volume III of the *Minnesota Studies in the Philosophy of Science*. Minneapolis: University of Minnesota Press. 1962.

fies the concept which is expressed by the name. To know this is to know that which is expressed by the existence statement formed from the name and the verb "to exist." Thus, in the case of common names, with which we shall primarily be concerned,

(31)   Ns exists
(32)   (Ex) x satisfies the criteria connoted by "N"
(33)   (Ex) x satisfies the concept of an N
(34)   (Ex) "N" names x

are different ways of making the same statement.

28. It will be clear from the above that I am committing myself to the view that *only those existentially quantified statements in which the quantified variable takes names of objects as its substituends have the force of existence statements.* The fact that its substituends are names of objects correlates the variable with a range of objects. On this account

(35)   (Ef) Socrates is f

would not be an existence statement; for the variable "f" does not in the above sense have a range of objects. The substituends for "f" are not *names,* but *predicates.* The variable "f" may indeed be said to have a *meaning range,* that is, the range expressed by

red, circular, wise, mortal, etc.

But *meanings unlike concepts are not objects,* and talk of meanings springs from the purely translational sense of 'means'.[17]

29. Notice that there is, of course, an object in the neighbourhood of "wise," but its name is "Wisdom," and it falls in the range of the concept variable "f-ness." Unlike (35), the closely related statement

(36)   (E f-ness) Socrates exemplifies f-ness

does satisfy the above mentioned necessary condition for being an existence statement. But if the preceding account is correct, expressions of the form "f-ness" name linguistic objects; and while, given that a language is rich enough to express the sense of

17. This point is elaborated in Chapter 8 [of original work, Ed.].

"Socrates exemplifies wisdom" as well as "Socrates is wise," these two statements are necessarily equivalent, it would be simply a mistake to assume that the quantified *object language* statement

(35)   (Ef) Socrates is f

has the sense of an existence statement asserting the existence of an extralinguistic abstract entity.

30. As a parenthetical remark it may be noted that it follows from the above that the Ramsey sentence of a theory is not an existence statement. (The Ramsey sentence of a formalized theory is, roughly, the sentence formed from the conjunction of its postulates by replacing all theoretical predicates by variables and prefixing the conjunction by these variables quantified 'existentially'.) Even though the Ramsey sentence does imply—given that we are willing to talk about the concepts expressed by theoretical terms—the existence of concepts or properties satisfying the conditions expressed by the postulates of the theory, the question whether these concepts or properties are *theoretical* or *observational* is simply the question whether constants substituted for the variables quantified in the Ramsey sentence can be construed, *salva veritate*, as belonging to the observational vocabulary.

31. According to the above analysis, then, to know that molecules exist is to know that

(37)   (Ex) x is ϕ_1 . . . ϕ_n

where being ϕ_1 . . . ϕ_n is a sufficient condition in the framework of the theory for somebody to satisfy the concept of a molecule. The question arises, under what circumstances can we be said to know this? Note that while (37) is a statement in the language of the theory, it need not be construed as either a postulate or a theorem of the theory. What the theory does is provide us with a licence to move from statements in the observation language asserting the existence of a certain physical state of affairs at a certain time and place to statements asserting the presence of a group of molecules at that time and place. To know that molecules exist is to be entitled to the observational premises, and to be entitled to the licence to move from this premise to the theoretical conclusion. To be entitled to this license is for the theory to be a good theory.

32. Carnap's distinction [18] between internal and external questions of existence is relevant here. For even if the question, "Are there molecules?" is one which cannot be answered without going outside the language of the theory in the narrow sense of this phrase, it *is* internal to the framework provided by the functioning of a theory as a theory. And as such it can be contrasted with the 'external' question, "Is there good reason to adopt the framework of molecules?"

III. Microtheoretical Explanation

33. It would seem, then, that if kinetic theory is a *good* theory, we are entitled to say that molecules exist. This confronts us with a classical puzzle. For, it would seem, we can also say that if our observation framework is a *good* one, we are entitled to say that horses, chairs, tables, etc., exist. Shall we then say that *both* tables and molecules exist? If we do, we are immediately faced with the problem as to how theoretical objects and observational objects "fit together in one universe." To use Eddington's well-worn example, instead of the one table with which pretheoretical discourse was content, we seem forced to recognize two tables of radically different kinds. Do they both *really* exist? Are they, perhaps, *really* the same table? If only one of them *really* exists, which?

34. It has frequently been suggested that a theory might be a *good* theory, and yet be *in principle* otiose. (Not otiose; *in principle* otiose.) By this is meant that theory might be known on general grounds to be the sort of thing which, in the very process of being perfected, generates a *substitute* which, in the limiting case of perfection, would serve all the scientific purposes which the perfected theory could serve. The idea is, in brief, that the cash value at each moment of a developing theory is a set of propositions in the observation framework known as *the observational consequences of the theory*, and that once we separate out the heuristic or 'pragmatic' role of a theory from its role in explanation, we see that the ob-

18. See his essay on "Empiricism, Semantics, and Ontology," printed as an appendix in his *Meaning and Necessity*, 2nd Edition, Chicago: University of Chicago Press, 1959.

servational consequences of an ideally successful theory would serve all the scientific purposes of the theory itself.

35. If we knew that theories of the kind we are considering were *in principle* otiose, we might well be inclined to say that there *really* are no such things as molecules, and even to abandon our habit of talking about theoretical expressions in semantical and quasi-semantical terms. We might refuse to say that theoretical terms express concepts, or name or denote objects; we might refuse to say that theoretical objects exist. And we might well put this by claiming that theoretical languages are *mere* calculational devices. This resolution of the initial puzzle has at least the merit of being neat and tidy. It seems to carve theoretical discourse at a joint, and to cut off a superfluous table with no loss of blood.

36. I shall argue that this is an illusion. But what is the alternative? One possible line of thought is based on the idea that perhaps the observational level of physical things (which includes one of the tables) has been mistakenly taken to be an 'absolute'. It points out that if the framework of physical things were in principle subject to discard, the way would be left open for the view that perhaps there is only one table after all; this time, however, the table construed in theoretical terms.

37. I think that this suggestion contains the germ of the solution to the puzzle; but only if it is developed in such a way as to free it from the misleading picture—the levels picture—which generates the puzzle. It is, I believe, a blind alley if, accepting this picture, it simply argues that the observational framework *is itself a theory*, and that the relation of the framework of microphysical theory to it is implicitly repeated in its relation to a more basic level—the level, say, of sense contents. For though this account might well enable one to dispense in principle with the physical objects which serve as mediating links between sense contents and molecules (the latter two being capable, in principle, of being directly connected), nevertheless we should still be left with *two* tables, a cloud of molecules on the one hand, and a pattern of actual and possible sense contents on the other.

38. My line will be that the true solution of our puzzle is to be found by rejecting as in paragraph 36 the unchallengeable status of the physical-thing framework, *without, however, construing this framework as a theory with respect to a mere basic level.* My strategy

will be to bring out the misleading and falsifying nature of the *levels* picture of theories. Thus I shall not be concerned, save incidentally and by implication, with the widespread view that the relation of physical-object discourse to sense-impression discourse is analogous to that of micro-theories in physics to the framework of physical things.

39. There are two main sources of the temptation to talk of theories in terms of levels: (i) In the case of microtheories, there is the difference of size between macro- and micro-objects. With respect to this I shall only comment that the entities postulated by a theory need not be smaller than the objects of which the behaviour is to be explained. Thus, it is logically possible that physical objects might be theoretically explained as singularities arising from the interference of waves of cosmic dimensions. (ii) The more important source of the plausibility of the *levels* picture is the fact that we not only explain *singular matters of empirical fact* in terms of *empirical generalizations;* we also, or so it seems, explain these generalizations themselves by means of *theories.* This way of putting it immediately suggests a hierarchy at the bottom of which are

Explained Nonexplainers,
the intermediate levels being
Explained Explainers
and the top consisting of
Unexplained Explainers

Now there is clearly *something* to this picture. But it is radically misleading if (*a*) it finds too simple—too simple in a sense to be given presently—a connection between explaining an explanandum and finding a defensible general proposition under which it can either be subsumed,[19] or from which it can be derived with or without the use of correspondence rules; (*b*) it is supposed that whereas in the observation framework inductive generalizations serve as principles of explanation for particular matters of fact, microtheoretical principles are principles of explanation *not (directly) for particular matters of fact in the observation framework but for the*

---

19. For a sustained critique of the subsumption picture of scientific explanation from a somewhat different point of view, see Michael Scriven's papers in Volumes I and II of the *Minnesota Studies in the Philosophy of Science,* and his unpublished doctoral dissertation (Oxford) on explanation.

inductive generalizations in this framework (the explaining being enacted with deriving the latter from the former) which in their turn serve as principles of explanation for particular matters of fact.[20]

40. This latter point is the heart of the matter; for to conceive of the *explananda* of theories as, simply, *empirical laws* and to *equate* theoretical explanation with the derivation of empirical laws from theoretical postulates by means of logic, mathematics, and correspondence rules is to sever the vital tie between theoretical principles and particular matters of fact in the framework of observation. Indeed, the idea that the aim of theories is to explain *not* particular matters of fact *but rather* inductive generalizations is nothing more nor less than the idea that theories are in principle dispensable. For to suppose that particular observable matters of fact are the proper *explananda* of inductive generalizations in the observation framework and of these only, is to suppose that, even though theoretical considerations may lead us to formulate new hypotheses in the observational framework for inductive testing and may lead us to modify, subject to inductive confirmation, such generalizations as have already received inductive support, the *conceptual framework* of the observation level is autonomous and immune from theoretical criticism.

41. The truth of the matter is that the idea that microtheories are designed to explain empirical laws and explain observational matters of fact only in the derivative sense that they explain explainers of the latter rests on the above-mentioned confusion between explanation and derivation. To avoid this confusion is to see that theories about observable things *do not "explain" empirical laws in the manner described, they explain empirical laws by explaining why observable things obey to the extent that they do, these empirical laws;* [21] that is, they explain why individual objects of various kinds

20. From a purely formal point of view, of course, one could derive ("explain") the observational consequence (C) of an observational antecedent (A) by using the theoretical theorem "$A_T \longrightarrow C_T$" and the correspondence rule "$A \longleftrightarrow A_T$" and "$C \longleftrightarrow C_T$" without using the inductive generalization "$A \longrightarrow C$." This, however, would only disguise the commitment to the autonomous or 'absolute' (*not* unrevisible) status of inductive generalizations in the observation framework.

21. The same is true in principle—though in a way which is methodologically more complex—of micro-microtheories about microtheoretical objects.

and in various circumstances in the observation framework behave in those ways in which it has been inductively established that they do behave. Roughly, it is because a gas is—in some sense of 'is'—a cloud of molecules which are behaving in certain theoretically defined ways, that it obeys the *empirical* Boyle–Charles law.

42. Furthermore, theories not only explain why observable things obey certain laws, they also explain why in certain respects their behaviour obeys no inductively confirmable generalization in the observation framework. This point can best be introduced by contriving an artificially simple example. It might, at a certain time, have been discovered that gold which has been put in *aqua regia* sometimes dissolves at one rate, sometimes at another, even though as far as can be observationally determined, the specimens and circumstances are identical. The microtheory of chemical reactions current at that time might admit of a simple modification to the effect that there are two structures of microentities each of which 'corresponds' to gold as an observational construct, but such that pure samples of one dissolve, under given conditions of pressure, temperature, etc., at a different rate from samples of the other. Such a modification of the theory would explain the observationally unpredicted variation in the rate of dissolution of gold by saying that samples of observational gold are mixtures of these two theoretical structures in various proportions, and have a rate of dissolution which varies with the proportion. Of course, if the correspondence rules of the theory enables one to derive observational criteria for distinguishing between observational golds of differing theoretical compositions, one would be in a position to replace the statement that gold dissolves in *aqua regia* sometimes at one rate, sometimes at another, by laws setting fixed rates for two *varieties* of observational gold and their mixtures. But it is by no means clear that the correspondence rules (together with the theory) *must* enable one to do this in order for the theory to be a good theory. The theory must, of course, explain why observational chemical substances do obey *some* laws, and the theoretical account of the variation in the rate at which gold dissolves in *aqua regia* must cohere with its general explanation of chemical reactions, and not simply postulate that there is an unspecified dimension of variation in the microstructure of gold which corresponds to this observed variation. But this is a

far cry from requiring that the theory lead to a confirmable set of empirical laws by which to replace the initial account of random variations.

43. Thus, microtheories not only explain why observational constructs obey inductive generalizations, they explain what, as far as the observational framework is concerned, is a random component in their behaviour, and, in the last analysis it is by doing the latter that microtheories establish their character as indispensable elements of scientific explanation and (as we shall see) as knowledge about what *really* exists. Here it is essential to note that in speaking of the departure from lawfulness of observational constructs I do not have in mind simply departure from all-or-none lawfulness ('strict universality'). Where microexplanation is called for, correct macroexplanation will turn out (to eyes sharpened by theoretical considerations) to be in terms of 'statistical' rather than strictly universal generalizations. But this is only the beginning of the story, for the distinctive feature of those domains where microexplanation is appropriate is that in an important sense such regularities as are available are not statistical *laws,* because they are unstable, and this instability is explained by the microtheory.

The logical point I am making can best be brought out by imagining a domain of inductive generalizations about observables to be idealized by discounting errors of measurement and other forms of experimental error. For once these elements in the 'statistics' have been discounted, our attention can turn to the logico-mathematical structure of these idealized statistical statements. And reflection makes clear that where microtheoretical explanation is to be appropriate, these statements must have (and this is a logical point) a mathematical structure which is not only compatible with, but calls for, an explanation in terms of 'microvariables' (and hence *micro-initial conditions:* the nonlawlike element adumbrated in the preceding paragraph) such as the microtheory provides. This point is but the converse of the familiar point that the irreducibly and lawfully statistical ensembles of quantum-mechanical theory are mathematically inconsistent with the assumption of hidden variables.

To sum up the above results, microtheories explain why inductive generalizations pertaining to a given domain *and any refinement of them within the conceptual framework of the observation language* are at best approximations to the truth. To this it is anticlimatic to add that theories explain why inductive generalizations hold only within certain boundary conditions, accounting for discontinuities which, as far as the observation framework is concerned, are brute facts.

44. My contention, then, is that the widespread picture of theories which equates theoretical explanation with the derivation of empirical laws is a mistake, a mistake which cannot be corrected by extending the term 'law' to include a spectrum of inductively established statistical uniformities ranging from 100 per cent to 50–50. Positively put, my contention is that theories explain laws by explaining why the objects of the domain in question obey the laws that they do to the extent that they do.

IV. Correspondence Rules Again

45. Suppose it to be granted that this contention is correct. What are its implications for the puzzles with which we began? The first point to be made is that if the basic schema of (micro-) theoretical explanation is,

> Molar objects of such and such kinds obey (approximately) such and such inductive generalizations because they *are* configurations of such and such theoretical entities.

then our puzzles are focused, as it were, into the single puzzle of the force of the italicized word "are," *Prima facie* it stands for identity, but how is this identity to be understood? Once again we are led to ask the methodological counterpart question, that is, what *is* a correspondence rule?

46. One possible but paradoxical line of thought would be that an effective microtheory for a certain domain of objects for which inductive generalizations exist is from the standpoint of a philosopher interested in the ontology of science, a framework which aspires to *replace* the observation framework. The observation framework would be construed as a poorer explanatory framework with a better one available to replace it. But thus boldly conceived, this replacement would involve dropping both the empirical generalizations and the individual observational facts to be explained by the theory, and would seem to throw out the baby with the bath. The

observation framework would be construed as a poorer explanatory framework with a better one available to replace it. Before we ask what could be meant by "replacing an observation framework by a theoretical framework," let us note one possible reaction to this suggestion. It might be granted that this is the sort of thing that is done when one theoretical framework is "reduced" to another, and that the notion of the replaceability of a microframework by a micro-microframework is a reasonable explanation of the force of such a statement as

> Ions behave as they do because they *are* such and such configurations of subatomic particles.

Yet the parallel explanation of the force of "are" where the identity relates not theoretical entities with other theoretical entities, but theoretical entities with observational entities might be ruled out of court. Once again we would have run up against the thesis of the inviolability of observation concepts on which the rejection of the replacement idea is ultimately grounded. This thesis, however, is false.

47. Nor is it satisfactory to interpret the proposal in paragraph 46 as follows: The framework of physical things is a candidate for replacement *on the ground that* it is actually a common sense theoretical framework, and *qua* theoretical framework may be replaced by another. For unless the conception of the framework of physical things as a replaceable explanatory framework goes hand in hand with an abandonment of the levels picture of explanation, it leads directly to the idea that below the level of physical-thing discourse is a level of observation in a stricter sense of this term, and of confirmable inductive generalizations pertaining to the entities thus observed.

48. The notion of such a level is a myth. The idea that sense contents exhibit a lawfulness which can be characterized without placing them in a context either of persons and physical things or of microneurological events is supported only by the conviction that it must be so so we are not to flaunt 'established truths' about meaning and explanation. Since my quarrel on this occasion is with these 'established truths', I shall not argue directly against the idea that there is an autonomous level of sense contents with respect to which

the framework of physical things plays a role analogous to that of a theory in the levels of explanation sense.

49. My answer to the question of paragraph 45 requires that we distinguish between two interpretations of the idea that the framework of physical things is an explanatory framework capable in principle of being replaced by a better explanatory framework. One of these interpretations is the view on which I have just been commenting. The alternative, in general terms, will clearly be a view according to which the framework of physical things is a replaceable theory-like structure in a sense that does not involve a commitment to a deeper 'level' of observation and explanation.

50. The groundwork for such a view has already been laid with the above rejection of identification of theoretical explanations with derivation. But what is a correspondence rule if it is *not* simply a device for deriving laws from theoretical postulates? We have seen that a correspondence rule is not a partial definition of theoretical terms by observation terms. Nor, obviously, is it a definition of observation terms as currently used by means of theoretical terms. But might it not be construed as a *redefinition* of observation terms? Such a redefinition would, of course, be a dead letter unless it were actually carried out in linguistic practice. And it is clear that to be fully carried out in any interesting sense, it would not be enough that sign designs which play the role of observation terms be borrowed for use in the theoretical language as the defined equivalents of theoretical expressions. For this would simply amount to making these sign designs ambiguous. In their new use they would no longer be *observation* terms. The force of the 'redefinition' must be such as to demand not only that the observation-sign design correlated with a given theoretical expression by syntactically interchangeable with the latter, *but that the latter be given the perceptual or observational role of the former so that the two expressions become synonymous by mutual readjustment*. And to this there is an obvious objection: *the meaningful use of theories simply does not require this usurpation of the observational role by theoretical expressions*. Correspondence rules thus understood would remain dead letters.

51. But if the above conception of correspondence rules as 'redefinitions' will not do as it stands, it is nevertheless in the neigh-

bourhood of the truth; for if correspondence rules cannot be regarded as implemented redefinitions, can they not be regarded as statements to the effect that certain redefinitions of observation terms would be in principle acceptable? This would be compatible with the fact that the redefinitions in question are implemented only in the syntactical dimension, no theoretical expressions actually acquiring the observational-perceptual roles they would have to have if they were to be synonyms of other expressions playing this role. This view has at least the merit of accounting for the peculiar character of correspondence rules as expressing more than a factual equivalent but less than an identity of sense. It would explain how theoretical complexes can be unobservable, yet 'really' identical with observable things.

52. On one classical interpretation, correspondence rules would appear in the material mode as statements to the effect that the same objects which have observational properties also have theoretical properties, thus identifying the denotation, but not the sense, of certain observational and theoretical expressions. On another classical interpretation, correspondence rules would appear in the material mode as asserting the coexistence of two sets of objects, one having observational properties, the other theoretical properties, thus identifying neither the denotation nor the sense of theoretical and observational expressions. According to the view I am proposing, correspondence rules would appear in the material mode as statements to the effect that the objects of the observational framework *do not really exist—there really are no such things.* They envisage the *abandonment* of a sense and its denotation.

53. If we put this by saying that to offer the theory is to claim that the theoretical language could beat the observation language *at its own game* without loss of scientific meaning, our anxieties are aroused. Would not something be *left out* if we taught ourselves to use the language of physical theory as a framework in terms of which to make our perceptual responses to the world? I do not have in mind the role played by our observational concepts in our practical life, our emotional and aesthetic responses. The repercussions here of radical conceptual changes such as we are envisaging would no doubt be tremendous. I have in mind the familiar question: would not the abandonment of the framework of physical things mean the abandonment of the *qualitative* aspects of the world?

54. To this *specific* question, of course, the answer is yes. But it would be a mistake to generalize and infer that *in general* the replacement of observation terms by theoretical constructs must "leave something out." Two points can be touched on briefly. (*a*) I have suggested elsewhere [22] that the sensible qualities of the common sense world, omitted by the physical theory of material things, might reappear in a new guise in the microtheory of sentient organisms. This claim would appear in the material mode as the claim that the sensible qualities of things *really* are a dimension of neural activity. (*b*) There is an obvious sense in which scientific theory cannot leave out qualities, or, for that matter, relations. Only the most pythagoreanizing philosopher of science would attempt to dispense with descriptive (that is, nonlogical) predicates in his formulation of the scientific picture of the world.

22. Chapter 1, section VI; also Chapter 3, section VII [of original work, Ed.].

Edward MacKinnon

Edward MacKinnon studied physics at Boston College, Harvar
University, and St. Louis University, where he received his Ph.D
His graduate studies in philosophy were done at Boston College
where he received an M.A., and at Yale University, where he spen
two years as a post-doctoral fellow. He also did graduate work i
theology and received the theological degree S.T.L. from Westo
College. A former professor at Boston College, he is now a pro
fessor of philosophy at California State University, Hayward. H
has written many articles on the philosophy of science, analyti
philosophy, and epistemology. His abiding concern with the inte
relation of science, philosophy, and theology is manifested in h
recent book, *Truth and Expression: The 1968 Hecker Lectures* (Pa
ramus, N.J.: Paulist/Newman Press, 1971).

Atomic Physics and Reality

The development of physics in the twentieth century centers around two great achievements, the theory of relativity and quantum theory. It is generally agreed that the latter is the most revolutionary and far-reaching theory in the history of physics since the days of Newton. Here, however, general agreement ends, for the proper interpretation to be given to quantum theory and its relation to reality has been a subject of continuing debate for over thirty years. The most widely accepted view, the so-called "orthodox interpretation" championed by Bohr and Heisenberg [1] represents a peculiar mixture of physics and philosophy. Their contention is: No one can consistently accept the facts of atomic physics and reject their interpretation of them; no one can logically accept their interpretation and reject its philosophical implications. Nevertheless, the philosophical implications have been found unacceptable by many philosophers of different schools. [2]

From *The Modern Schoolman*, 38 (1960): 37–59, by permission of the editor.

1. Bohr first explained his views at a physics conference at Como, Italy, in September of 1927. This talk is reprinted in his *Atomic Theory and the Description of Nature: Four Essays with an Introductory Survey* (Cambridge: Univ. Press, 1934). One of Heisenberg's earliest explanations is contained in *The Physical Principles of the Quantum Theory* (Chicago: Univ. of Chicago Press, 1930). The latest and most complete exposition of each of these authors' views may be found in Niels Bohr, "Discussion with Einstein on Epistemological Problems in Atomic Physics," *Albert Einstein: Philosopher-Scientist* (Evanston, Ill.: Library of Living Philosophers, 1949), pp. 199–241; Werner Heisenberg, "The Development of the Interpretation of the Quantum Theory," *Niels Bohr and the Development of Physics*, ed. W. Pauli (New York: McGraw-Hill Book Co., 1955), pp. 12–30; Werner Heisenberg, *Physics and Philosophy: The Revolution in Modern Science* ("World Perspectives," Vol. XIX; New York: Harper & Bros., 1958). Chapter viii of this last book is a slightly less technical version of the preceding account by Heisenberg.

2. Eva Cassirer, "Methodology and Quantum Physics," *British Journal for*

Before accepting or rejecting any part of this physics-philosophy one must know which aspects of it are clearly established, which are doubtful, and which are gratuitous assumptions. This we intend to indicate in the present paper by outlining the historical development of the theory and the reasons which have been adduced for and against it. The theory itself, the so-called "Copenhagen interpretation," need not be considered in complete detail since it has been adequately treated elsewhere.[3] The related philosophical problems will be considered in a subsequent article. In both articles we shall attempt to separate the philosophical problems from their physical basis to the degree that the matter permits, since we believe that such a separation is a prerequisite for clarification of this involved problem.

Early Development

Quantum theory has developed through a series of jumps, experimental and theoretical breakthroughs, which established new plateaus. Each jump was followed by a detailed examination of the new plateau and eventually, when it proved incomplete, a prepara-

---

the Philosophy of Science, VIII (1958), 334–41, summed up the reaction of the participants in the Second International Conference of the Philosophy of Science (held in Zurich, 1954): "Complementarity was almost generally rejected as being unhelpful." A summary of many of the philosophical objections to this theory may be found in Mario Bunge, "Strife About Complementarity," *British Journal for the Philosophy of Science,* VI (1955), 1–12; 141–54.

3. For concise, not overtechnical explanations of this doctrine see Norwood Russell Hanson, "Copenhagen Interpretation of Quantum Theory," *American Journal of Physics,* XXVII (1959), 1–15; Enrico Cantore, "Philosophy in Atomic Physics: Complementarity," *The Modern Schoolman,* XXXIV (1957); Philipp Frank, *Philosophy of Science: The Link between Science and Philosophy* (Englewood Cliffs, N.J.: Prentice-Hall, 1957) chaps. ix and x. A more technical treatment is given by Hans Reichenbach, *Philosophical Foundations of Quantum Mechanics* (Berkeley: Univ. of California Press, 1944), Pt. 1. A summary of this problem and his solution is contained in his posthumous work, *The Direction of Time,* ed. Maria Reichenbach (Berkeley: Univ. of California Press, 1956), sec. 25.

tion for a new jump. The critical breakthrough which precipitated the conflicts on the nature of fundamental reality came with the simultaneous development of quantum mechanics and wave mechanics, which occurred, roughly, between 1924 and 1927. However, this conflict can only be understood in the light of the development which preceded it.[4]

The earliest stages in the development of quantum theory—Planck's explanation of black body radiation (1900) and Einstein's explanation of the photoelectric effect (1905)—introduced an apparent contradiction into the description of light. Maxwell and others had successfully unified and explained all the known data on the basis of a wave theory of light. Planck's theory, as interpreted by Einstein, pictured light as a stream of photons and on this corpuscular basis explained new and otherwise intractable data. The further development of quantum theory by Debye, Bohr, and Sommerfeld did nothing to resolve this conflict.

Louis de Broglie's answer to this dilemma was a radical one.[5] Rather than resolve the duality, he extended it to mass particles (electrons and protons) by postulating that every fundamental "particle" was composed of a corpuscle plus an associated wave. Through this theory he was able to give an explanation of the apparently arbitrary principle (quantization of orbits) which had been used heuristically by Bohr to explain the structure and some of the properties of the atom. Schrödinger (1926) proved that the requisite conditions for the wave particle duality postulated by de-Broglie could not be satisfied within the atom. He eliminated the corpuscular aspect and assumed that all the known and knowable particles were of a wave nature. On this assumption he developed

4. A. d'Abro, *The Rise of the New Physics* (2 vols.; New York: Dover Pubns., 1951), Vol. II, gives a detailed semitechnical history of this development. A brief, popular, yet authoritative history of atomic theory may be found in Werner Heisenberg, *Nuclear Physics,* trans. F. Gaynor and A. von-Zepplin (London: Methuen, 1953). A historical account which stresses the problem of interpretation may be found in Max Born, *Physics in My Generation: A Selection of Papers* (London: Pergamon Press, 1956).

5. Louis deBroglie explains the development of his own theory in *The Revolution in Physics: A Non-mathematical Survey of Quanta,* trans. Ralph W. Niemeyer (New York: Noonday Press, 1953).

wave mechanics, which proved an unprecedented success in solving the problems that perplexed his contemporaries.

Heisenberg, Born, and Jordan simultaneously and independently of Schrödinger developed quantum (or matrix) mechanics on a completely different basis.[6] Emulating and extending the precedent set by Bohr, Heisenberg totally rejected all physical models of atomic reality, such as waves and particles, and developed a mathematical formalism which used only observable quantities, such as frequencies and polarizations. His cumbersome mathematical methods did not facilitate the solution of practical problems as did the more flexible wave mechanics. Yet the two theories did not seem to lead to any contradictory conclusions.

In the summer of 1926 Bohr, as the elder statesman of atomic physics, invited both Schrödinger and Heisenberg to Copenhagen to discuss the new developments. Much to Schrödinger's displeasure, Bohr proved that the simple wave picture could not even explain the original work of Planck. Heisenberg's collaboration with Bohr proved more satisfactory. As Heisenberg later recalled:

> During the months following these discussions and intensive study of all questions concerning the interpretation of quantum theory in Copenhagen finally led to a complete and, as many physicists believe, satisfactory clarification of the situation. But it was not a solution which one could easily accept. I remember discussions with Bohr which went through many hours till very late at night and ended almost in despair; and when at the end of the discussion I went alone for a walk in the neighboring park I repeated to myself again and again the question: Can nature possibly be as absurd as it seemed to us in these atomic experiments?[7]

What were these experiments which seemed to demonstrate that nature was neurotic? A detailed explanation of some of them has been given by Bohr.[8] Here we will give a sketchy explanation of the simplest of them.

---

6. Heisenberg gives an interesting nontechnical account of the genesis his own ideas in *The Physicist's Conception of Nature,* trans. Arnold J. Pomerans (London: Hutchinson, 1958), pp. 51–71.

7. Heisenberg, *Physics and Philosophy,* p. 42.

8. Niels Bohr in *Albert Einstein,* pp. 210–30.

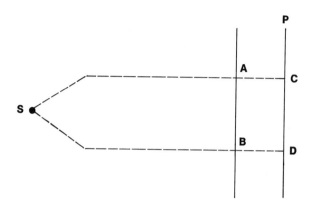

A source, S, ideally an infinite distance away, projects electrons through the slits A and B and on to the photographic plate, P, where they are recorded by the spot produced. We consider first the results when slit B is closed. The spots have a symmetrical distribution about a point of maximum intensity, C, which is on a line with A and S. Are these electrons behaving as waves or particles? The laws proper to waves indicate that a wave passing through A would spread and could not produce a sharp point interaction. The occurrence of such sharp interactions could be seen most clearly if the beam were weakened until one electron came through at a time and produced a single spot on the plate. If slit A is closed and B opened, a similar intensity pattern is found centered about the point D. Accordingly, only a particle interpretation seems tenable.

Now both slits are opened. We would expect a distribution of spots which would be the sum of the two preceding ones. This is not observed. Instead the photographic plate would reveal a series of bright and dark lines, an interference pattern similar to the beats produced by sound waves. If one attempts to explain this observation on a particle basis he must assume that the behavior of a particle passing through slit B depends on whether or not slit A is opened, an assumption which clearly seems to be unreasonable.

The difficulty illustrated by this experiment has been explained rather precisely by Reichenbach.[9] Calling "phenomena" the atomic

9. Reichenbach, *Philosophical Foundations*, pp. 1–42.

facts which may be inferred rather directly—here the emission of electrons and their subsequent absorption at the plate—and "interphenomena" any explanation of the behavior of the particles between the coincidences which constitute the phenomena, he summarizes the two conclusions drawn from the experimental data. (1) One of the two proposed interphenomena (particles and waves) can explain *any* possible experiment. (2) Neither of the two interphenomena can be used to explain *all* possible experiments without introducing causal anomalies.

To the nonphysicist, this dilemma may seem to be of trivial significance. To the atomic physicist it is quite serious. The ultimate goal, the ideal towards which atomic physics had been striving, was an explanation of fundamental particles in terms of basic principles and of composite bodies in terms of the properties and activities of the particles which constitute these bodies. Accordingly, these fundamental particles, which are still but dimly understood, were thought to represent reality at its most basic level. As de-Broglie expressed it:

> Nevertheless, it seems to me that one thing can be asserted: despite the importance and the extent of the progress accomplished by physics in the last centuries, as long as the physicists were unaware of the existence of quanta, they were unable to comprehend anything of the profound nature of physical phenomena for, without quanta there would be neither light nor matter and, if one may paraphrase the Gospels, it can be said that without them "was not anything made that was made." [10]

Hence the magnitude of the crisis. As long as quantum theory remained a thing of rags and patches nature was not understood. A unified view was needed before the new plateau could be properly explored.

The Copenhagen interpretation of quantum theory born of the collaboration of Bohr and Heisenberg developed along two different but related lines. Bohr tried to work the new pictures obtained by wave mechanics into the interpretation of the theory in an acceptable way. This was not an attempt to build a constructural model of an atom. Rather, it was motivated by his realization that, how

10. DeBroglie, *Revolution in Physics,* p. 14.

ever far the phenomena to be explained may transcend the scope of classical physical explanation, the account of all evidence must be expressed in classical terms. The successful development of quantum theory on the basis of evidence obtained from classical instruments indicated that the notions proper to classical physics must have some validity when applied to the phenomena measured by these instruments.

Bohr's analysis of actual and possible experiments led to two conclusions which concern us here. The first is the *indivisibility of atomic phenomena*. This conclusion is based on the realization that the atomic system to be studied is interacting on an atomic level with the experimental apparatus and thus destroying the clear-cut distinction between the process of observation and the thing observed—or, in Bohr's terms, between the subject and the object. The whole experiment must be considered as an epistemologically irreducible unit. The significance of this "indivisibility" principle may be seen by considering the double slit experiment previously described. The question, "Which of the two slits does the electron pass through?" is an attempt to subdivide the phenomena observed. To ask such a question in a physically meaningful way one must change the experimental set-up; for example, by closing one slit. However, the question is considered meaningless when asked of the experiment performed with both slits open. In this way the previous difficulty is solved by simply legislating it out of existence with, according to Bohr, nature itself presiding as the legislator.

The second principle, *complementarity,* is closely related to the first. The dilemmas discussed at Copenhagen revealed an essential ambiguity in any attempt to apply conventional physical attributes to atomic objects. For example, to apply the concept of momentum one must use an experimental arrangement which measures momentum exchange. This momentum exchange prohibits precise localization of the interaction which supplies the experimenter with his information. In a similar way one can measure position only by destroying the possibility of measuring momentum. Accordingly, the two concepts, momentum and position, are to be considered as complementary concepts. Though both are needed, no possible experiment can give a precise determination of both simultaneously.

The spontaneous objection to this doctrine is that it confuses the possibilities of measurement with the properties of the things to be measured. Though the precise measurement of a particle's momentum may destroy the possibility of measuring its location, one instinctively feels that before the experiment was performed the particle did have a location. It must have been somewhere. This difficulty is, in a peculiarly inverted way, the cornerstone of Bohr's theory. All knowledge of atomic behavior is rooted in experimental data. This data allows no possibility of using the concept "location" in a meaningful way without becoming involved in causal anomalies; for example, the way a particle goes through slit A depends on whether or not slit B is open. In terms of Bohr's principles this discussion of location—albeit an unmeasurable location—is a conceptual attempt to divide an indivisible phenomenon. The apparent contradiction involved in speaking of a particle without location is due to the use of the term "particle" beyond the limits set by the principle of complementarity. Under the circumstances, it would be preferable to use the term complementary to "particle"; that is, "wave." For wave phenomena—for example, a sound—localization has, at best, a rather vague meaning.

Heisenberg's contribution to the Copenhagen interpretation was of a more abstract nature. Bohr met the paradox of apparently contradictory experimental results by imposing severe restriction on the language in which these experiments could be described. Heisenberg faced the problem of how to pass from an experimentally given situation to its mathematical representation. He solved this problem by reversing it. In conjunction with Pauli he developed a mathematical formalism (the representation of states by vectors in Hilbert space) and introduced the hypothesis that only those states which could fit into the formalism could occur in nature. This assumption led directly to the now famous "indeterminacy principle." [11] This principle focused Bohr's doctrine of complementarity into a sharp mathematical form which specified the maximum accuracy obtainable in the simultaneous measurement of position and momentum or other similarly related complementary concepts. Again, the paradoxes are legislated out of existence, but now by mathematical rather than linguistic laws.

11. The "experimental" derivation of the indeterminacy principle given by Heisenberg in *Physical Principles* has often led to the simplified explanation that the indeterminacy results from the fact that the experiment disturbs reality. Bohr has objected to this oversimplification on the grounds that introduces a distinction between the observation and the reality observed.

The mathematical synthesis which Heisenberg strived for was furthered by other developments. Chief among these was the re-interpretation of Schrödinger's wave mechanics. Born (1926) showed how these waves could be interpreted as determining a probability distribution.[12] At the same time, Schrödinger and Eckart independently demonstrated that the mathematical formulation of wave mechanics is equivalent to the more involved matrix methods. Finally, Heisenberg and Kennard sounded the death knell of the original wave picture proposed by Schrödinger by proving that wave packets diffused in time and could not be considered to behave like small particles.[13] These separate elements were gathered together and harmonized in the monumental work of Dirac.[14]

The result of these labors is a complete consistent system which is capable, at least in principle, of answering all questions within its domain. By this statement we certainly do not mean that all questions concerning fundamental particles, for example, have been answered or are likely to be answered in the immediate future. What it does mean is that any proposed answer must, according to the Copenhagen school, be consistent with the formalism of quantum theory and be interpreted in the light of its principles. Accordingly, subsequent generations of physicists could content themselves with learning quantum mechanics and the rules for applying it, while almost completely ignoring the "philosophical" problems of interpretation. This, it seems, is what the overwhelming majority have done. As Einstein noted in this connection ". . . only those who have successfully wrestled with the problematic situations of their own age can have a deep insight into those situations." [15] Accordingly, the philosophy of the Copenhagen interpretation and of quantum theory in general will be extracted from the writings of those who wrestled with this problematic situation.

12. Born's theory is summarized in his Nobel Prize lecture, "Statistical Interpretation of Quantum Mechanics," reprinted in *Physics in My Generation*, pp. 177–88.

13. Schrödinger still holds a modified form of his original wave theory. His present views are explained in "Are There Quantum Jumps?" *British Journal for the Philosophy of Science*, III (1953), 109–34, 233–42, and in *Scientific American*, CLXXXIX (Sept. 1953), 52. The "orthodox" answer to Schrödinger's objections is given in Born's article, "The Interpretation of Quantum Mechanics," *Physics in My Generation*, pp. 140–50.

14. A. M. Dirac, *The Principles of Quantum Mechanics* (3rd ed.; Oxford: Oxford Univ. Press, 1957).

15. Albert Einstein, "Reply to Criticisms," *Albert Einstein*, p. 665.

Philosophical Problems

That the "Copenhagen Interpretation" is intended to be a philosophy as well as a physics seems clear from the statements of its authors. As Heisenberg summarized it:

> What was born in Copenhagen in 1927 was not only an unambiguous prescription for the interpretation of experiments but also a language in which one spoke about Nature on the atomic scale, and in so far a part of philosophy. . . . This was not, however, the language of one of the traditional philosophies, positivism, materialism, or idealism; it was different in content, although it included elements from all these systems of thought.[16]

This philosophy differs from traditional philosophies in its mode of expression and development as well as in its content. It is almost inextricably intertwined with the physics from which it springs and thus presents, for the philosopher of more traditional views, an elusiveness which is most frustrating. This interconnection is not unintentional. Behind it lies the contention that no one can logically accept the present quantum theory and deny the philosophical implications which flow from it.[17] The enormous success of quantum theory points to the conclusion that this sublimated physics is a scientifically proven philosophy. Under these circumstances any commentator who attempts to separate the philosophy from its physical basis runs the risk of distorting the mind of the authors of the theory. Nevertheless the risk must be taken.

Perhaps a general, though certainly oversimplified, statement of this philosophy could be that the "quantum philosophy" consists in translating the restrictions imposed upon physical theory into epistemological and metaphysical statements. The epistemological problem is the decisive one, since their metaphysics ultimately consists in a denial of the possibility of metaphysics. In a grossly oversimplified form the epistemological argument might run as fol-

16. Heisenberg, *Niels Bohr*, p. 16.
17. This claim is most clearly seen in Heisenberg's refutation of those who claim to accept the physics of his theory but reject the philosophy (*ibid.*, pp. 17–23).

lows. All knowledge comes through the senses—and the men under consideration (Bohr, Heisenberg, Pauli, Dirac, and to a lesser degree, Born) will admit of no knowledge essentially distinct from sense knowledge. The greatest refinement of sense knowledge is had in the experimental data of modern physics. This data manifests an intrinsic limit in the amount and type of knowledge which can be obtained. Hence it is literally meaningless to speak, in any way, of anything beyond these limits. This view has been succinctly expressed by Dirac: "Only questions about the results of experiments have a real significance and it is only such questions that theoretical physics has to consider." [18] A similar summary was given by Bohr:

> The extent to which an unambiguous meaning can be attached to such an expression as "physical reality" cannot of course be deduced from *a priori* philosophical conceptions but . . . must be founded on a direct appeal to experiments and measurements.[19]

This epistemology, a logical extension of sensism, plays a determining role in the "metaphysics" of complementarity. Only phenomena can be known; nothing more can be said to exist. Bohr has repeatedly explained that we cannot attribute autonomous physical reality (that is, a reality independent of the experimenter) to objects on the atomic scale.[20] Recently, Heisenberg has expounded this doctrine in a more philosophical tone. He distinguishes three types of realism—practical realism, dogmatic realism, and metaphysical realism—by the following criterion.[21] One "objectivates" a statement if he claims that its content does not depend upon the conditions under which it can be verified. Practical realism assumes that there are statements that can be objectivated and that, in fact, the largest part of daily experience consists in such statements. Dogmatic realism, the working philosophy of prequantum physics, claims that there are no statements concerning the material world that cannot

18. Dirac, *Principles of Quantum Mechanics,* p. 5.
19. Niels Bohr, "Can Quantum Mechanical Description of Physical Reality Be Considered Complete?" *The Physical Review,* XLVIII (1935), 696.
20. Bunge, *British Journal for the Philosophy of Science,* I, 3, n. 1, gives a bibliography of the various articles in which Bohr defends this opinion.
21. Heisenberg, *Physics and Philosophy,* p. 81.

be objectivated. Finally, metaphysical realism goes one step further by saying that "the things really exist." Dogmatic realism falls with the quantum paradoxes. Metaphysical realism is equated with Descartes's doctrine on the *res extensa* and then simply dismissed by an appeal to Locke, Berkeley, and Hume.

Elsewhere, this doctrine of realism and objectivity is applied to atomic problems.[22] For an atomic system to be objective it must be completely isolated from the rest of the world. Any connection would introduce submicroscopic disturbances which, to an unknown degree, destroy the objectivity of the property in question. A measurement supplies just such a disturbing connection. Hence, a completely objective system can never be known. Is it real? The answer to this question must be sought in the sources of knowledge from which the term "real" gets its meaning. The real is that which can be known by the senses and, by extension, verifiable concepts which may be derived from sense knowledge. Such concepts can be verified of an atomic system only by destroying its objectivity; that is, by performing a measurement. Then the system becomes real but is no longer objective. In summary, an atomic system may be either objective (and "potential") or real, but not objectively real. Only macroscopic systems, which are not appreciably changed by observation, may be considered objectively real.

The metaphysical realist may be inclined to dismiss this doctrine as simply absurd. Yet two claims may be made in its behalf. First, the experimental data, the quantum theory that explains it, and the epistemology and metaphysics sketched above form a consistent system. Secondly, the quantum theory, in which this philosophy is embedded, has proved extremely successful. Any future theories must contain the present theory, at least as a limiting form. For these reasons this "quantum philosophy" seems destined to perdure. It must be taken seriously.

What criticisms should be brought against this theory? Rather than attempt an independent answer to the question we shall consider the chief criticisms which have been offered through the years and the responses which the defenders of the Copenhagen interpretation have proffered.

22. Heisenberg, *Niels Bohr,* pp. 25-28.

Conflict over Complementarity

The objections to this theory, like the theory itself, represent a mixture of philosophy and physics put forth by men who, for the most part, are professional physicists and amateur philosophers. Here, however, the relation between the two is even more confused. Many of the objections, implicitly based on what Heisenberg would call "metaphysical realism," were elaborated by men who explicitly denied the possibility of metaphysics. In their criticisms, the language of physics hovers over the problem of the real existence of the objects of physics. We shall present the objections, chiefly those of Einstein, and then try to extract the philosophical residue.

The trial by fire for the new theory came with the Solvay physics conference held in Brussels in October, 1927.[23] The older generation of physicists who had inaugurated the quantum theory (Planck, Einstein, vonLaue, Lorentz) strongly objected to the new proposed interpretation on the ground that it violated the very nature of a physical theory. Implicit in the methodology of classical physics, including the theory of relativity, were two requirements which any theory must satisfy. First among these is causality in the physicists's sense; that is, a subsequent state is determined by a preceding state. Second among the requirements traditionally imposed, especially on mechanical theories, is that such a theory must explain all phenomena as relations between objects existing in space and time. By discussing various possible experiments, Einstein, as the leader of the opposition, showed that the new theories, wave mechanics and quantum mechanics, could not satisfy these standards.

Bohr's reply, which was more fully developed later, was that since these standards could not be met, the standards themselves must change to meet the new theory. Causality and space-time localization should be considered complementary and mutually exclusive aspects of atomic phenomena. There are exact laws of a causal form (the time-dependent Schrödinger equation) and there are space-time observations. But it is not possible to establish a one-to-one

23. The discussion between Bohr and Einstein at this conference is summarized by Bohr in *Albert Einstein,* pp. 211–24.

correspondence between the two. The relation is merely statistical. Implicit in this defense is a radically new idea of the very nature of scientific explanation. Science does not explain things or even give laws for the activities and properties of things. Rather, it serves to correlate the various observations which can be made. This distinction was of little significance as long as it could be held, at least in principle, that observation did not disturb the system being observed. In atomic physics the observed system and the observation fused into an indivisible unit, a phenomenon, and the distinction became crucial. Two years later Bohr explained this theory of physics a little more clearly:

> We meet here in a new light the old truth that in our description of nature the purpose is not to disclose the real essence of the phenomena but only to track down, so far as it is possible, relations between the manifold aspects of our experience.[24]

At the same conference the wave-particle theory made a last, rather desperate, stand.[25] DeBroglie had developed a "pilot-wave" theory which, he hoped, might unify the wave and particle aspects of quantum phenomena in a conceptual model capable of explaining the observed data. Pauli proved that this new theory violated the firmly established principles of special relativity. For this reason, and also because of mathematical difficulties in the theory, deBroglie abandoned the attempt to form unified models and capitulated to the Copenhagen school.

Einstein's criticism took on a sharper edge in the 1930 Solvay conference.[26] He argued that when the exigencies of the special theory of relativity were properly considered, the indeterminacy principle and the doctrine of complementarity do not constitute an absolute limit in the knowability of atomic phenomena. He proposed an ideal thought experiment, a light-tight box containing radiation and a clock which controlled a shutter capable of releasing a single photon. Through the clock mechanism the *time* at which

24. Niels Bohr, *Atomic Theory*, p. 18.
25. See deBroglie, *Revolution in Physics*, chap. x, sec. 6.
26. This is summarized by Bohr in *Albert Einstein*, pp. 224–30. A dramatic nontechnical account of this debate is given by George Gamow, "The Principle of Uncertainty," *Scientific American*, CXCVIII (Jan. 1958), 51–57.

the energy of the system changed could be precisely determined. By an ideally precise weighing of the box before and after the experiment, coupled with the famous relativity formula, $E = mc^2$, the *energy* change could also be precisely determined.[27] Yet the Copenhagen interpretation insisted that time and energy were complementary variables governed by an indeterminacy relation. They could not be precisely determined simultaneously.

After a hectic sleepless night Bohr returned to the conference with a detailed answer. By invoking the "principle of equivalence," the fundamental principle of Einstein's own general theory of relativity, he showed that an energy change in the system would change the rate at which the clock measured time. The resulting indeterminacy in time, coupled with the indeterminacy that, as he proved, must be found in the weighing, yielded the relation exactly as the indeterminacy principle required.

Since this time the claim of the Copenhagen school that their interpretation was able to present a consistent theory which could, at least in principle, explain all the then known data has not been seriously challenged to my knowledge. As Einstein himself later summarized the situation:

> It must be admitted that the new theoretical conception owes its origin not to any flight of fancy but to the compelling force of the facts of experience. All attempts to represent the particle and wave features displayed in the phenomena of light and matter, by direct course to a space-time model, have so far ended in failure. And Heisenberg has convincingly shown, from an empirical point of view, any decision as to a rigorously deterministic structure of nature is definitely ruled out because of the atomistic structure of our experimental apparatus. Thus it is probably out of the question that any future knowledge can compel physics again to relinquish our present statistical theoretical foundation in favor of a deterministic one which would deal directly with physical reality.[28]

This admission of Einstein does not mean, however, that the Einstein-Bohr conflict ceased. It merely changed form, coming

27. It is of interest to note that Einstein obtained both of his relativistic formulas discussed here by similar thought experiments.
28. Albert Einstein, *Out of My Later Years* (New York: Philosophical Lib., 1950), p. 109.

slightly closer to philosophy. The Copenhagen school insisted that the limitations in the "knowability" of phenomena were determined by the very nature of the phenomena. Accordingly, the ψ-function, together with its statistical interpretation, offered an absolute limit to atomic knowledge. In this sense it could be considered exhaustive. This conclusion, proper to the formalism of a physical theory, was sublimated into the philosophical proposition that there was no atomic reality beyond that represented in the ψ-function. Einstein, together with Podolsky and Rosen, felt that the ψ-function was incomplete.[29] Certain aspects of physical reality, they insisted, must be considered real and yet could not be included in the present theory. As might be expected, they offered a criterion of physical reality based on a correlation with physical theory:

> If, without in any way disturbing a system, we can predict with certainty (i.e., with probability equal to unity) the value of a physical quantity, then there exists an element of physical reality corresponding to this physical quantity.

By applying this criterion of reality to yet another ideal experiment, the authors tried to demonstrate that both position and momentum could be simultaneously *real*, though they were not simultaneously measurable. They considered two separate systems, I and II, which interacted for a time and then separated. By arguments which need not be recounted, they showed that it should be possible to measure the position of I and then, by means of the product wave function ψI,II' determine the position of II, *without in any way disturbing II*. By their criterion of reality, the position of II is real. Similarly, the momentum of II could be determined by measuring the momentum of I, again without disturbing II. Hence, they concluded, both position and momentum are simultaneously real in spite of the fact that they are not simultaneously measurable.

Bohr's reply [30] translated the general requirements of the EPR

29. A. Einstein, B. Podolsky and N. Rosen, "Can Quantum-Mechanical Description of Reality Be Considered Complete?" *Physical Review*, XLVI (1935), 777–80.

30. Niels Bohr, "Can Quantum-Mechanical Description of Reality Be Considered Complete?" *Physical Review, LXVIII* (1935), 606–702. A more theoretical explanation of the EPR experiment in terms of ensemble theory has been given by Armand Siegel and Norbert Wiener, "Theory of Measurement

experiment into a system which, in principle, was amenable to precise measurement. An examination of the possible measurements which could be performed on the system revealed that any measurement of the position of I would destroy all possibility of determining the momentum of II and vice versa. He concluded that their criterion of reality was essentially ambiguous when applied to atomic systems. They spoke of measuring I without in any way influencing II, whereas the very conditions that made a measurement of I possible necessarily exerted an influence on II. In spite of the separation of the two systems, the experiment as a whole must be considered one indivisible phenomenon.

A study of both papers leaves one with a rather frustrated feeling. Either paper, considered by itself, seems correct. When viewed together they do not quite meet, in spite of the apparent clear clash. Both are trying to establish the relationship between physical theory and the things observed; that is, physical reality. Yet each would define both terms of the relationship differently. Bohr, as we have seen, considers physical theory to be essentially a means of correlating observations, while "physical reality" is practically equivalent to the resulting correlation. Einstein's conception of the nature of physical theory is, by his own admission, derived from his own work in general relativity and field theory.[31] By this standard any theory which is not a representation of things in space and time must be considered incomplete, though it may represent the maximum utilization of available information.

His ideas on "physical reality" are of more interest to us.[32] Though he finds it impossible to give a critical justification of his stand—without exhuming the dead bones of metaphysics—he holds for a distinction between an "objective" and a "subjective" factor in knowledge, even on the atomic level. If one can admit such an objective element—that is, a *Ding an sich* which is really out there—the philos-

in Differential-Space Quantum Theory," *Physical Review,* CI (1956), 429–32. The alternative explanation proposed by W. Furry (*Physical Review,* [1936], 393, 476) that there is no correlation between the separated systems seems to have been disproved experimentally. See D. Bohm and Y. Aharonov, "Discussion of Experimental Proof for the Paradox of Einstein, Rosen, and Podolsky," *Physical Review,* CVIII (1957), 1070–76.

31. Einstein, "Autobiographical Notes," *Albert Einstein,* p. 89.
32. Einstein, "Reply to Criticisms," *ibid.,* p. 673.

ophy of Bohr and Heisenberg, a repudiation of such objective reality, must be incorrect.

Plausible as this stand may have seemed on intuitive grounds, Einstein found himself in extreme difficulty every time he tried to present a critical defense of it. Each of his many attempts to construct verifiable statements about the "objective" element in atomic reality had met with frustration. Any attempt to construct metaphysical statements about atomic reality was precluded by his epistemology, the doctrine that concepts and propositions get meaning only through connection with sense experience.[33] There remained only the hope, poignantly expressed, that future physicists would be able to verify his intuition of reality:

> Some physicists, among them myself, can not believe that we must abandon, actually and forever, the idea of direct representation of physical reality in space and time; or that we must accept the view that events in nature are analogous to a game of chance. It is open to every man to choose the direction of his striving; and also every man may draw comfort from Lessing's fine saying, that the search for truth is more precious than its possession.[34]

What conclusions may be drawn from this long, involved controversy? One hesitates to arbitrate this war of the giants; yet certain points seem to be fairly well established. First among them is the realization that the Copenhagen interpretation forms a complete consistent system capable of integrating, at least in principle, all of the atomic facts which are now known or which are likely to be discovered by an extension of the present theoretical and experimental methods. Many physicists share Einstein's hope that a better system may be found. Few, if any, would deny the success of the present system at least as a minimal interpretation. Secondly, the peculiar philosophy of Bohr, Heisenberg, and others, a sort of "semi-idealism" which holds that macroscopic objects are real but that submicroscopic objects become real only through being observed, is not based merely on a repudiation of metaphysics. It relies on the fact that any attempt to make meaningful statements about the behavior of an atomic system between observations must lead to contradictions. Thirdly, the problem of the nature of atomic reality

33. *Ibid.,* p. 13.
34. Einstein, *Out of My Later Years,* p. 110.

is, at least in the literal sense, a "meta-physical" problem, since the problem of existence is beyond the scope of physics. This is indicated by the fact that both sides agree on the pertinent data and the mathematical formalism which describes it, while they are in complete disagreement on the reality described. Obviously, such a metaphysical problem cannot be solved by those who deny the possibility of metaphysics or by those whose philosophy is simply a linear extrapolation from physics. Finally, any metaphysical realist who treats this problem must be prepared to render a critical justification of his privileged information, a justification which physics proved incapable of rendering.

Before concluding this survey we must mention, at least briefly, the recent flurry of proposed alternative interpretations of quantum theory.[35] Chief among them, the source and inspiration of the others, is Bohm's interpretation of quantum mechanims in terms of hidden variables.[36] Through extended discussions with Einstein, Bohm became dissatisfied with the Copenhagen interpretation and began a critical re-examination of the foundations of quantum theory. He discovered that a suitable substitution could split the Schrödinger equation into two equations, one of which could be interpreted as describing the trajectory of a particle provided the conjugate equation is interpreted as an equation governing the behavior of a new unobserved quantum potential energy.[37] A similar idea had been proposed earlier by deBroglie and abandoned because it seemed to

35. Heisenberg, *Niels Bohr,* gives a brief outline and refutation of each of the theories that disagrees with his own. A synthesis of the work done on the "causal interpretation" of quantum theory together with a complete (as of 1957) bibliography may be found in H. Freistadt, "The Causal Formulation of Quantum Mechanics of Particles (The Theory of deBroglie, Bohm, and Takabayasi)," *Nuovo Cimento Suppl.* (Ser. 10), V (1957), 1–70.

36. David Bohm, "A Suggested Interpretation of the Quantum Theory in Terms of 'Hidden Variables,'" I, *Physical Review,* LXXXV (1952), 166–79; II, *ibid.,* 180–93; "Reply to a Criticism of a Causal Re-interpretation of the Quantum Theory," *ibid.,* LXXXVII (1952), 389–390; "Proof that Probability Density Approaches $/\psi/^2$ in Causal Interpretation of the Quantum Theory," *ibid.,* LXXXIX (1953), 458–66.

37. The substitution $\psi = R \exp(2 \pi iS/h)$ is made in the Schrödinger equation. The resulting equation for S may be interpreted as a Hamilton-Jacobi equation, provided the equation for R governs the behavior of a new quantum potential energy, U. DeBroglie had developed similar equations in 1927 but abandoned them because of difficulties in their mathematical formulation. Bohm's new theory of measurement seems to solve, or at least cover over, these difficulties. See Bohm II, Appendix B (*Physical Review,* LXXXV, 180–93).

lead to inconsistencies. By carrying the "hidden variable" theory to its logical conclusion and introducing a new theory of measurement, Bohm demonstrated a way in which these apparent inconsistencies could be reconciled with the observed data.[38]

Rather than extend this article by a discussion of inappropriate technical details, we will, rather arbitrarily, present our own view on the present status of Bohm's theory. There are two points to be considered. First, the question, Is the Copenhagen interpretation the only possible interpretation of quantum theory? Bohm's detailed treatment of this point seems to show that any such assumption is unwarranted. That is, the assumption that there can be no more to atomic reality than is included in the formalism of quantum theory is based on the same positivistic limitation of scientific theory to explicit evidence that led Mach to deny the existence of atoms. Bohm, at least in his later works, presents his own hidden variable theory as an illustration of a possible alternative explanation of quanta. By its very construction and the interpretative postulates introduced, it is mathematically equivalent to the standard theory.[39] That such a theory can succeed at all seems to be proof enough that the Copenhagen interpretation is not the uniquely possible explanation of the facts.[40] On the other hand the fact that the new theory is self-consistent and mathematically equivalent to the established theory does not guarantee its reasonableness as a physical theory.

The second point, clearly a matter of opinion, is the success of Bohm's theory precisely as a physical theory. Because of the arbitrary foundation on which it rests, the purely gratuitous assumptions

38. David Bohm, *Causality and Chance in Modern Physics* (New York: Van Nostrand, 1957), chap. iii, gives rather convincing arguments to prove that the Copenhagen interpretation is not the uniquely possible interpretation of quantum mechanics.

39. Heisenberg, *Niels Bohr,* claims that Bohm's theory may be considered to be the Copenhagen interpretation expressed in a different language. Such a statement can be defended only on strictly positivistic grounds. It really sidesteps the main point of the new interpretation, its causal determinism.

40. We have deliberately omitted a mention of von Neumann's famous proof that the Copenhagen interpretation is unique, since his assumptions do not seem to apply to the hidden variables postulated by Bohm and others. This is discussed in Bohm, II sect. 9. See I. I. Zinnes, "Hidden Variables in Quantum Mechanics," *American Journal of Physics,* XXVI (1958), 1–4, for an explanation of von Neumann's theorem and its assumptions.

introduced, the highly implausible conclusions that result from it, and the complete lack of corroborating evidence we must conclude that his theory, in its present form, is unacceptable.[41] Bohm, and the other "causal theorists" keep promising new improved versions which, they hope, may contribute to a clarification of the difficulties in the present quantum field theories. Until and unless such promises are fulfilled, the Copenhagen interpretation will remain the "orthodox interpretation" of quantum theory.

Conclusion

The Copenhagen interpretation is an interpretation of quantum theory developed by Bohr and Heisenberg and extended by Born, Pauli, Dirac, and others. In spite of a long history of formidable opposition, it has been generally accepted as a self-consistent, physical, reasonable way of interpreting the known data on atomic and subatomic phenomena. Its essence, as a *physical* theory, involves a reinterpretation of the purpose and program of physics. Formerly, physics was expected to give a causal explanation of the behavior of objects chiefly in terms of space and time variables. The new physics uses these variables, but in a different way, as a means of correlating the observations that can be made.

For macroscopic objects the distinction is inconsequential; for submicroscopic events it is critical. Here the very observation—that

41. Heisenberg rejected this theory on the ground that it destroys the symmetry of the quantum equations, which he considers an essential part of the quantum theory. The implausible conclusions which we had in mind are first the quantum potential itself. As Bohm and Aharanov (*Physical Review*, CVIII, 1070–76), used it in explaining the ERP paradox the quantum potential between two photons is strong enough to determine the polarization state at a distance of several feet. It should be noted that this "explanation" violates the most essential condition of the original experiment, the requirement that a measurement in System I does not affect System II. Secondly, the proposed instantaneous interaction violates one of the basic postulates of special relativity. Bohm's answer, that the interaction cannot transmit a signal, does not seem to be consistent with the causal interpretation, since the quantum interaction signals the polarization state of one of the two photons. Finally, his theory leads to results which do not seem physically reasonable. The velocity of the unobserved particle varies from zero (for states with zero angular momentum) to infinity (for certain types of double slit experiments). However, the basic objection to the theory is the one mentioned in the text; it is a completely gratuitous theory which has no experimental verification.

is, the phenomenon observed considered as an irreducible unit—is the thing which science must explain. Any attempt to go beyond this by introducing a distinction between the observation and the object observed inevitably leads to causal anomalies when one tries to make meaningful (that is, capable of being verified, at least in principle) statements about the behavior of the object. Hence, Bohr concluded, any attempt to construct precise conceptual models of objects must be abandoned as a matter of principle. However, imprecise models—for example, particles and waves—may be used within certain limits. These "pictures" are allowed and to some extent necessary, not to slake a metaphysical craving but to supply a means for applying the ordinary concepts and language on which understanding depends. From this understanding a complete quantum theory has been developed. Consequently, post-Copenhagen physicists can concentrate on the mathematical formalism and its applications, and completely prescind from the semiphilosophical problems of interpretation.

The *philosophical* conclusions drawn from this theory, and apparently underwritten by its overwhelming success, are really nothing but a restatement of the physical principles in philosophical language. If one had to pin a label on this philosophy, it might be classified as a semi-idealistic metaphysics coupled to a positivistic theory of knowledge. That is, beyond the atomic phenomena there is nothing else. No meaningful way to speak of atomic objects or "things in themselves" can ever be found. These philosophical conclusions rest on two submerged but necessary assumptions.

The first assumption is that the Copenhagen interpretation represents an inherent limitation in knowledge which future generations of physicists will never be able to by-pass. Bohm and others have shown that this is an assumption and not a necessary consequence of the theory. The history of science teaches that few prohibitions limiting future scientists to the then current limitations have proved successful. However, it must be admitted that nothing in the present development of physics indicates that a return to a deterministic physics is likely. The opposite, a further departure from classical ideals, seems much more probable.

The second assumption is that metaphysical statements about atomic phenomena are meaningless. Apart from the general tenor of positivism, two reasons may be given for this rejection. The first is a misunderstanding, partial or total, of the nature of metaphysics.[42] Metaphysical statements, according to this misinterpretation, are statements that go beyond the available evidence. An example would be a statement of the behavior of "observed" particles between observations. The second and more basic reason is epistemological. The Copenhagen theory, as Bohr sees it, represents a rational utilization of *all* possibilities of unambiguous interpretation of measurement. Accordingly, he envisions no grounds on which further pertinent statements of any sort can be based. Implicit in this, to be sure, is a rejection of any distinction between fundamentally different ways of knowing. The metaphysical realist, the man who, according to Heisenberg's thumbnail sketch, holds that "things are really there," must be prepared to give some sort of an answer to such elementary and obvious questions as: "What is really there?" and "How do you know?"

42. Henry Margenau, *The Nature of Physical Reality: A Philosophy of Modern Physics* (New York: McGraw-Hill, 1950), p. 12, has given a vivid description of the confusion of physics with metaphysics that is sometimes encountered: "Many reputable scientists have joined the ranks of the exterminator brigade, which goes noisily abroad chasing metaphysical bats out of scientific belfries. They are a useful crowd, for what they exterminate is rarely metaphysics—it is usually bad physics."

Mario Bunge

Mario Bunge was born in Buenos Aires and educated in physics and philosophy. He has taught philosophy in Argentina, in Europe, and in various American universities. At present he is a professor of philosophy at McGill University in Montreal. Probably no other contemporary philosopher of science has manifested as critical and as continuing a concern for the problem of scientific realism as Bunge. Even in treating such technical problems as the axiomatic foundations of quantum mechanics and the validity of the Copenhagen interpretation, developed by Bohr and Heisenberg as a means of explaining the distinctive features of quantum mechanics, Bunge is concerned with a critical development of scientific realism. The following essay is a summary of his views on the logical and epistemological problems involved in scientific realism. He has also developed detailed analyses of causality, of the requirement of simplicity in scientific explanation, and on the metascientific questions arising from scientific inquiry.

Physics and Reality

1. Introduction

Every science is about some class of objects or other. In particular, physics deals with sets of physical objects: theoretical physics is supposed to represent certain features of objects of a kind—namely, physical systems—and experimental physics assumes the task of testing such theoretical representations. And those objects that are the concern—or, as we shall say, the intended referents—of physical theory are, *ex-hypothesi*, self-existent: they are not mind-dependent. True, some of them, such as the transuranians, might not have come into being without human action guided by physical theory; others, such as the magnetic monopoles, may be no more than fictions. And every idea concerning physical objects of a kind, whether or not it is an adequate idea, is no more and no less than an idea. Furthermore, no such idea is ever a photographic description of its intended referent but a hypothetical incomplete and symbolic representation of it. Yet the point at issue is that physical theory intends to refer ultimately to *real* objects and, moreover, in the most *objective* (i.e., subject—detached or operator—invariant) and *true* (adequate) possible way.

What follows spells out the preceding platitudes and attempts to analyze some traits of physical theories that often obscure its intended real reference, objectivity, and partial truth.

From *Dialectica: Revue Internationale de Philosophie de la connaissance*, 20 (1966), 2:174–195, by permission of the editor. Numbers in brackets refer to References, p. 257–58.

2. Reference [1]

When talking of temperatures we intend to characterize the thermal states of some physical system, such as a body or a radiation field. In this case the referent of our assertions is a physical system or perhaps a class of physical systems. This reference is tacit rather than explicit: it is taken for granted since it is suggested by the context. Yet by omitting to point out the objective reference we may forget that physical concepts aim at representing properties of physical systems. The same holds for every constant (nonaccidental) relation among physical variables, i.e. for every law of physics. Thus, when writing down an equation of state we intend this formula to refer to some physical system or, rather, to be about an arbitrary member of a certain class of physical systems. The same holds, *a fortiori,* for systems of law statements, i.e. theories.

The objective reference can be made more precise by mathematization; yet at the same time this refinement procedure, if misinterpreted, will becloud reference. In fact, the aim of mathematization in physics is to represent things and their properties on a conceptual plane, dealing henceforth with these deputies rather than with their constituencies. Thus, what is usually focused on in the mathematical representative of a physical variable is not the whole concept but just the numerical part (s) of it. Take again the temperature concept: what we insert in a thermodynamical law statement is not the whole temperature concept but the numerical variable θ occurring in the propositional function $t(\sigma/s) = \theta$, which is short for 'The temperature of system σ reckoned in the scale–*cum*–unit system s equals θ.' The reason for fixing s, leaving the object variable σ aside, and seizing on the numerical component θ, is clear: only mathematical concepts can be subjected to arithmetical computation and θ is, of the whole temperature concept, precisely that ingredient capable of coming under the sway of arithmetic. But the momentary focusing on one of the ingredients of the temperature concept should not make us forget that temperature is not a numerical variable but a function mapping a certain set, built in part out of the set of physical systems, onto a set of numbers. (Briefly: let Σ denote the set of physical systems, S the set of scale *cum* unit systems, and $\Theta \subset R$ a subset of the real numbers. Then T maps the cartesian product of Σ and S onto Θ, i.e. $T: \Sigma \times S \to \Theta$. Whereas every $\sigma \epsilon \Sigma$ is supposed to be in the external world, S and Θ are constructs.)

(Something similar holds for any of the more complex physical variables. For example, the full quantum-mechanical representative of the linear momentum should not be written p or even p_{op} but rather $p_{op}(\sigma)$—which in fact we do whenever we intend to refer to the momenta of the individual components of an actual assembly of quantum-mechanical systems. In the present case the physical property is not represented by an ordinary function, but this is beside the point: the intended objective referent, here denoted by σ, is usually made clear by the context and this is why, whenever it consists of a single system, it can be dropped during the computations. But it must be kept in mind under pain of losing sight of physical meanings and, consequently, of rendering physical tests pointless.)

The philosopher, and sometimes even the physicist, may overlook the objective referent physical variables intend to point to, tending to think of temperature, or of any other physical concept, as a symbol in itself, and likewise of a set of equations as exhausting a physical theory. An analysis of physical variables restores their intended objective reference, by distinguishing the object variable (s) σ from the remaining variables involved in the conceptual representation of a physical property. Oddly enough, while a little analysis may chase realism away, a stronger dose of it will bring us closer to the view that physics aims at accounting for some aspects of reality: that physics is concerned with physical objects rather than with either mathematical structures or with our perceptions.

3. Direct and Indirect Reference [2]

The objective reference a physical variable is assumed to have must be distinguished from a direct—e.g. pictorial—representation of physical objects. Take temperature once more: as Mach recognized, the temperature concept is a brain-child of ours even though it was introduced in order to symbolize objective thermal states. Moreover, since there can be infinitely many temperature scales and units, there is some arbitrariness in our choice of any of them. (In other words, there is a number-to-body correspondence since there is at least one possible physical system to which any given temperature value $\theta \epsilon \Theta$ can be assigned. But there is no converse, body-to-number correspondence (function) unless a scale-*cum*-unit system is specified, since only then can we assign a single number to at least one physical system. In short, as we had before, $T: \Sigma \times S \rightarrow \Theta$.) The naive realist will emphasize the reference of every possible temperature concept to the set of all possible physical systems, whereas the conventionalist will underline the arbitrariness of the choice of scale and unit and, from this arbitrariness, he will conclude to the absence of objective reference.

We must grant each contendor a point. The numerical value of the temperature of a given system not being unique, a photographic representation of thermal states is out of the question. But once a scale has been chosen, the favored temperature function will represent in its own way the set of possible thermal states of physical systems. After all, not even photographers are required to photograph their subjects always from the same angle. Moreover, while the choice of a given scale-*cum*-unit system is conventional, it is not wholly arbitrary. Thus, the absolute scale is preferable to others for most purposes; 1) because it is independent of the peculiar behavior of any thermometric substance, and therefore one step removed from human limitations, and 2) because it fits best the statistical interpretations of thermodynamics. That is, the convention whereby the Kelvin scale and unit are nowadays preferred is grounded rather than capricious. The reason absolute temperature values are independent of any particular real substance and of any human operator is that the concept was tailored to specify the thermal states of the ideal gas. Such states are unreal because the ideal gas itself is a construct. Yet this construct is not a fiction: the ideal gas is supposed to be a theoretical schematization or model of a real gas. The various equations of state of the ideal gas that have been proposed so far refer immediately to this conceptual model rather than to any real gas.

Physics is not a game: a physical model, however unintuitive, is always a conceptual sketch of some object assumed to be out there. That this existence hypothesis can turn out to be false is beside the point. The point at issue, in the controversy over realism and objectivism, is that the physicist invents some key concepts (e.g., "temperature") which he somehow assigns to physical objects (e.g., thermal states of bodies). This concept-physical object correlation is partly stated in the interpretation rules assigning a physical meaning to the given symbols (see Sections 4 and 7). The ideal or theoretical models are supposed to represent, in a more or less symbolic—i.e. indirect and conventional—way, and to a certain approximation, some features of the constitution and behavior of physical systems. Every such model is part of at least one physical theory. (What can be regarded as being essentially the same model serves occasionally different theories: thus all electromagnetic field theories, whether or not they use potentials, and whether or not they are linear, share essentially the same field model even though they differ in the properties they assign to it, just as all direct inter-particle action theories share the black-box model.)

The above may be rephrased in the following negative way: no physical theory directly depicts or portrays a physical system. Firstly, because every theory is built with concepts, not with images, and those concepts, far from being empirical (e.g., observational), are full-fledged constructs, i.e. transobservational concepts such as "mass," "charge," "temperature," and "field strength." Secondly, because such key concepts are comparatively few in every theory and consequently they refer, if at all, to just a few chosen aspects of physical objects (those assumed to be important), rather than to the real physical system in all detail, i.e. such as it would be known to a supremely attentive and acute observer [3]. In short, every physical theory must be, as Duhem remarked, both symbolic and incomplete—from which it does not follow that it lacks an existential import or objective reference.

In fact, every physical theory does intend to represent an arbitrary member of a class of physical systems. It certainly does it symbolically and in a simplified way rather than iconically and completely; still, it does aim to represent such a real existent. Otherwise the very problem of building a theory would not be posed. And whenever such an attempt fails badly, the theory is changed or abandoned: the picture it provides is recognized as being either unfaithful (false) or operator-dependent (subjective).

When speaking of the reference of a physical idea (variable, statement, theory) we must therefore distinguish direct from indirect reference. Every physical construct refers *directly* to some *theoretical* model or other, i.e. to some ideal schematization embodied in a theory assumed to account, even modestly, for a physical system of a kind. The same construct refers therefore *indirectly* to some aspects of such a physical object (see Figure 1). Thus, the mediate referent

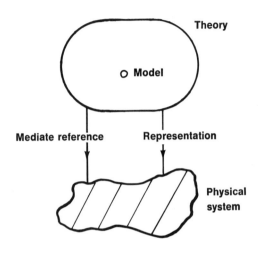

Figure 1   Objective reference: a correspondence between a model and real object.

of classical thermodynamics (thermostatics) is any reasonably insulated physical system, which is represented as a fluid in a perfectly closed container the fluid-*cum*-walls being (the theory's model). Of course there are no such systems in nature, with the exception of the universe as a whole. But any system contained in a calorimeter satisfying approximately the closure condition can be said to be a mediate referent of thermostatics.

In plain words: physics intends to represent reality but it does it in a hypothetical, roundabout, and partial way. (In more detail: a physical theory T is about a definite conceptual class U — the universe of discourse of T. U. corresponds to—but is not—some part Σ of reality. The phrase "T refers immediately to U" means that the formulas of T hold, by stipulation, for any element of U, i.e. for the model. And the expression "T refers mediately to Σ" means that U is assumed to correspond to $\Sigma \epsilon R$, whether or not the formulas of T remain true when the members of U occurring in them are replaced by the corresponding members of Σ. If the theory not only refers to facts but in addition does it in a verisimilar way, so much the better. In this case, i.e. for a good theory, we have for every σ in Σ and every x in U

$$[\Sigma \ \epsilon \ R \text{ and } \Sigma \neq \phi \text{ and } U \neq \phi \text{ and } X(x) \ \epsilon \ T]$$
$$\rightarrow (\exists \epsilon) \ [\epsilon \geqslant 0 \text{ and } X(x) \underset{\epsilon}{=} X \ (\sigma)]$$

where $\underset{\epsilon}{=}$ stands for 'equal (or equivalent) to within the error ϵ'. U is a definite class since it is determined by the predicates and assumptions of the theory. On the other hand the mediate referent Σ is a nondefinite class: its specification being incomplete, any number of borderline cases can crop up. This will force us to distinguish two kinds of meaning rules in Section 7. But before doing this let the operator enter the stage.

4. Interpretations: Objective and Operational [4,5]

A symbol occurring in a physical theory is either purely formal (logical or mathematical) or it can be assigned some nonformal (factual) meaning. In turn, a factually meaningful sign occurring in the language of a physical theory can be assigned an objective and/or an operational interpretation. Thus i may symbolyze the (perhaps unknown) intensity of an electric current while no ammeter is measuring it; but the same sign may, on a different occasion, stand for the measured value of "the same" current—after allowance has been made for the fraction of current dissipated in the measuring device. In brief, i may be read objectively and/or operationally. (The individual values need not coincide in both cases: for one thing, the measured values will always be affected by some experimental error that is absent from the theoretical calculations.) On the other hand basic magnitudes such as field potentials, lagrangians, or ψ-functions, can be assigned an objective but not an operational meaning. Only some of the concepts built with their assistance can be attached an operational interpretation

(e.g., $p = \delta L / \delta q$ in the case of a charged particle in a magnetic field.)

In general, we shall say that a symbol is given an *objective interpretation* when a rule is laid down which assigns to that sign a physical object (thing, property, event, process), whether or not the object is under observation and whether or not the reference relation is actually satisfied by the sign-object couple. And we shall say that a symbol is endowed with an *operational interpretation* (not definition!) if a correspondance is stipulated between the symbol and results of actual or feasible operations designed to observe or measure "the same" property of "the same" object. (The quotation marks intend to suggest that the object may change as a result of such empirical operations.) There will be no harm in assigning interpretations of those two kinds to one and the same symbol as long as the distinction is not blurred.

Every operationally meaningful symbol can also be interpreted in an objective way. The converse is not true: objective meanings are more fundamental and universal than operational ones. Thus the predicates "free particle," "intensity of a light wave in vacuum," and "stationary state of an atom" are all assumed to have objective referents although they cannot be assigned operational interpretations. The reason for this impossibility is plain: measurements, and particularly so atomic measurements, involve a coupling between the mensurandum and some aspects of an experimental set-up, whereby some of the properties of the initially free object are altered. And the reason for regarding objective meanings as more fundamental than operational ones is this: anything that is made up of a physical object interacting with a piece of apparatus constitutes a third, more inclusive, physical system entitled to be studied as such and consequently calling for objectively meaningful symbols. Only the *test* of theoretical statements concerning this wider system will require the interpretation of some of the terms occurring in them in terms of laboratory operations.

What has just been said holds for quantum physics as well as for classical physics. The chief difference is that in classical physics the theories of measurement are advanced enough to enable us to calculate (predict) the disturbances introduced by specified empirical operations, whereas no such mature theories exist for the atomic and subatomic domains. In other words, in classical physics we can account for the objective difference between a natural system and another object of the same kind interacting with our physical means of observation: the interaction is incorporated into the law statements and the result of the calculation can be checked empirically for the system under measurement. If the prediction calculated by means of the theory of measurement is borne out by measurement, the theory concerning the natural object is regarded as confirmed (not as verified). On the other hand, the quantum-mechanical theory of measurement is not yet capable of yielding similar results. (We shall here disregard the mutually incompatible claims that the available theory does its full job and that no such theory is conceivable because the object-subject interaction is ultimately unanalyzable, i.e. irrational.) In any case, there are physical and epistemological differences between a natural system and a measured one, and correspondingly between the objective and the operational interpretation of a physical symbol.

Our semantical distinction is meaningless in the context of the operationalist philosophy of physics. So much the worse for this philosophy, since that distinction is in fact employed in physics, notwithstanding the strenuous efforts made to reduce every idea to perceptions and operations taking place in a conceptual vacuum. Consider the quantum theory, sometimes regarded as a child of operationalism. One usually starts by stating some problem concerning what is assumed to be an autonomously existing object, such as an arbitrary helium atom. This atom may be assumed to be in its fundamental state, but this we will be unable to put in evidence unless we first excite it to some other energy level. This problem is not of the theorist's concern; on the other hand the experimentalist is forced to use some theory concerning the unobservable stationary states and the possible transitions among them: in order to produce those transitions he must have some idea about the required energy. Moreover, no experiment proper need be made in order to check the theoretical calculations in this case, as nature provides us with what we hypothesize as being helium atoms in various excited states randomly decaying to lower states. The corresponding measurements will alter none of the properties of our atoms, since the measurements consist in collecting and analyzing the light spontaneously emitted by the atoms. On top of this, the atoms themselves may well be located beyond the reach of our laboratory: they may dwell, say, somewhere in the Crab nebula. In short, it is not true that every quantum-mechanical calculation concerns a system coupled to a measurement set-up, much less to the observer's mind; and it is not true that every measurement relevant to quantum theory produces disturbances, let alone wholly unpredictable ones.

The circumstance that in every fundamental physical theory we handle some natural object or other rather than systems subject to severe test conditions, is tacitly acknowledged when posing a typical problem in theoretical physics. In fact, in such a problem statement only variables referring to the system under study will ordinarily occur. (Thus, when using a hamiltonian theory we start by writing down, i.e. by hypothesizing, the hamiltonian corresponding to our physical system, or rather to a schematic model of it, and go on to look for an eigensolution of the hamiltonian. In particular, this solution may be time-independent, thus representing a stationary state, which is unobservable. Usually no perturbation representing the non-hypothesized interaction of our system with a phony experimental set-up will occur in the hamiltonian: the latter will contain just the position, time, and momentum coordinates of the supposedly autonomously existing system—or rather of a sketch of it, such as an oscillator.)

Yet many physicists, misled by what used to be a fashionable philosophy, smuggle into the theorems something that was missing in the initial assumptions—namely, a measurement device and eventually even its operator, with his thoughts and his unpredictable intentions. This is how Heisenberg's relations are frequently interpreted, although no symbol representing measurement operations —let alone mental events—occurs in the axioms from which they are derived. Another example: the possible energy states of a free helium atom are, in the same vein, interpreted as the possible results of energy measurements—which measurements would involve perturbations that had not been assumed to begin with, i.e. when writing down the Schrödinger equation for the helium atom. This equation contains no variable referring to the structure and behavior of the phony measurement device, and it is only to comply with a philosophy accepted in advance that the theorems are interpreted in a way unwarranted by the initial assumptions. In brief, the solution to the original problem is somehow interpreted as the solution to an entirely different problem—a shift designed to introduce the Operator into the innermost recesses of nature. Our quantum philosopher-physicist is thereby making use of a privilege formerly reserved to theologians: namely, that of 'concluding' from one statement to a second statement referring to an altogether disjoint universe of discourse. Let us take a closer look at this commonly practised yet ill-studied strategy.

5. Conceptual Unity—and How to Violate It in Q. M.

The move we have just discussed exemplifies an illegitimate shift of meaning whereby symbols initially assigned an objective meaning are suddenly interpreted operationally. This manœuvre is carried out without caring whether such a reinterpretation is at all possible. If no such reinterpretation is justified, the 'conclusions' drawn in this way remain equally unwarranted. Such meaning shifts characterize the usual (phenomenalistic, operationalistic, idealistic) interpretations of both the quantum theory of "particles" and the quantum field theory, which are thereby rendered what shall be called *semantically inconsistent*. Since the trivial yet important concept of semantical consistency, or conceptual unity, does not seem to have been analyzed, a brief excursus will be convenient at this point. A fuller treatment is given elsewhere [6].

It is a desideratum for every theory to possess both formal and semantical unity. The former consists in the system's logical togetherness, i.e. in being a hypothetico-deductive system rather than an arbitrary heap of formulas. The *semantical consistency* or conceptual unit of a factual theory boils down to this: the system must be about some (nonempty) class which, far from being an arbitrary collection, is characterized by certain mutually related properties. Let us give a more precise characterization of semantical consistency.

To begin with, the conceptual unity of a theory requires a common reference of its formulas to some collection of objects. In the case of a physical theory, this collection is not an arbitrary set but a natural (nonarbitrary) class of physical objects. The class of objects the theory refers to is the latter's *universe of discourse*. Thus, the universe of discourse of fluid mechanics is the class of all fluids: the theory assigns the latter certain properties every one of which it represents by a certain predicate. It is taken for granted that the universe of discourse or reference set is not void, and it is hypothesized that the members of it can be paired to external objects in such a way that the theory holds at least approximately. Such a reference to external objects can be indirect and even false (see Section 3), but some reference to physical objects is always presupposed in a physical theory, and this is why it is called *physical* rather than, say, psychological. Theories carry the names of their ultimate referents, even if the latter turn out not to exist; thus, a theory referring to hyperons will be called a hyperon theory. On the other hand, a (meta) statement such as "The statements of quantum mechanics are not about autonomous physical systems but about our knowledge" is not only incomplete—as it fails to point out the object of such a knowledge—but it is a tacit claim that the quantum theory is not a physical theory.

Unity of reference is necessary yet not sufficient for a theory to attain full conceptual unity. A second factor of semantical consistency is that the predicates of the theory belong to a single family— briefly, that they be *semantically homogeneous*. Thus a physical theory will contain only predicates designating physical objects (systems, properties, events, processes). On the other hand a statement like "The wave-function propagates in (configuration) space and summarizes the observer's experimental information" mixes physical and information-theoretical predicates, thus muddling the distinction between a symbol, ψ, the state of the physical system it is supposed to represent (in a devious way indeed), and the bits of empirical information that may have been used in hypothesizing its precise form. (On top of this the statement suggests the false idea that ψ can be built directly out of data, without making hypotheses concerning the constituents of the systems, their interactions, and the law which ψ obeys.)

A common universe of discourse U, and a semantically homogeneous family P of predicates, are necessary but still insufficient to ensure the conceptual unity of a theory. One needs, in addition, the prohibition to smuggle into the theory predicates foreign to the field covered by the theory. This third condition, which may be called the requirement of *semantical closure*, may be stated thus: The predicates of the theory shall be just those which occur in the predicate base and in the definitions of the theory. Were it not for this semantical requirement, formal logic would consecrate the semantically dirty trick of validly deducing theorems containing concepts not occurring among those at the base of the theory. In fact, the rule of addition—t entails t or u—would allow us to surreptitiously adjoin, to any theorem t of a given theory, a statement u violating the condition of semantical homogeneity, by virtue of containing some concept not germane to the predicate base initially envisaged. This expansion of the initial base might go as far as to change the original universe of discourse in an arbitrary way: we might start by talking about atoms as physical objects and end up by talking about our behavior, whether or not in connection with atoms.

(Moreover, the outsider u might be a thoroughly untestable proposition, e.g., an *ad hoc* hypothesis designed to save the theory from empirical refutation. As if this were not enough, t or u is logically weaker than the genuine theorem t, hence easier to confirm—as easy as we may care. Finally, since any theorem of the theory may be found to be wanting in some respect or other, its negation will allow us to detach the unwanted u, i.e. to keep the stranger as the sole survivor of scientific criticism. Even if u were testable, and furthermore beyond practical doubt, the trick would defeat the purpose of the theorist, who is concerned with accounting for the referents of his theorem t, not for those of u. The rule of semantical closure aims at preventing such a manoeuvre. It can be shown [6] that the previously mentioned requirement of semantical homogeneity is not sufficient to rule that trick out.)

A fourth condition of semantical consistency is that the key concepts (basic predicates) of the theory hang together, by being fairly distributed among the initial assumptions of the theory. This may be called the condition of *conceptual connectedness*. It can be stated in a more precise way and it can be shown [6] that unity of reference and conceptual connectedness are necessary for attaining formal unity, as deducibility relations can only be established among formulas sharing certain key predicates, among which U stands out.

Every factual theory should possess both formal and conceptual unity, if only for methodological reasons, such as avoiding cheap confirmation. Unfortunately some physical theories, though formally (logically and mathematically) consistent, are semantically inconsistent, as they violate some or all of the first three requirements of conceptual unity, i.e. unity of reference, semantical homogeneity, and semantical closure. As anticipated in the preceding section, this is the case with the usual interpretations of the quantum theory: sometimes the eagerness to secure testability, at other times the wish to avoid an ontological commitment, and at other the hope of reviving subjectivistic philosophies, motivate the attempts to smuggle the Operator into a domain to which it did not belong to begin with; finally the Operator grabs the handle and the physical object is gone —from physics!

Yet no semantically consistent interpretation of quantum mechanics in purely operational terms (preparation, measurement, experiment) is available. In other words, there exists no consistent Copenhagen formulation of quantum mechanics: the physical interpretation proposed by that school does not match all the basic formulas of the theory (see Section 4). Moreover, the present formalism of quantum mechanics does not seem to allow for it, as such a theory would have to include from the start the consideration of measurement set-ups, dispensing with every term which, like "free particle," can be assigned an objective meaning but not an operational one. (Since the fundamental theory would have to refer only to objects under measurement or experiment: 1) it would make no sense to distinguish in the hamiltonian, the lagrangiagn, or any other source expression, the free part from the one representing the interaction of the system with a macroscopic set-up, and 2) we would be deprived from the guidance of classical mechanics in guessing the adequate quantum-mechanical hamiltonian —a hard task as things stand.) Even if a semantically consistent interpretation of quantum mechanics in the spirit of operationalism were eventually formulated: 1) it would not be a strictly physical theory but rather a psychological or an operations-research one, and 2) it would be inapplicable to objects which, like the atoms in Andromeda, we can afford to leave alone, without the assistance of the Operator.

In sum, *we lack a semantically consistent quantum theory,* whether in operationalist, idealistic, or realistic terms. And we should want to formulate a semantically consistent and thoroughly physical quantum theory that might in principle be applied to either an autonomous object or, *mutatis mutandis,* to a system under experimental control, and moreover such that the latter be handled as just a physical system of a special kind rather than as a mind-body *compositum.* Physicalism is a narrow ontology: granted; but it does work for the physical universe, and every retreat from physicalism in the realm of physics is a return to prescientific anthropocentrism. Why misplace the human mind: is it not a system of functions of certain bodies made up of atoms, and is it not enough to credit the human mind with the invention of theories, the planning of tests, and the interpretation of the latter's outcomes?

(We should note in passing that a realistic interpretation of the quantum theory does not require renouncing its present fundamental stochastic character. In other words, one does not need to introduce extra hidden variables in order to restore objectivity in the quantum domain: the hidden variables are already there. Only, they are miscalled *observables,* although no one can seriously claim that any of the fundamental variables of the quantum theories are strictly, i.e. directly observable or measurable. Hidden variables, in the sense of nonstochastic—nonfluctuating, scatter-less—magnitudes, are sufficient, yet not necessary, to produce a nonstochastic theory resembling classical dynamics. But such neoclassical concepts are very likely insufficient, and certainly dispensable, to build the much-needed semantically consistent and thoroughly physical interpretation[s] of the quantum-mechanical formalism. One should not mix the problems of reality and objectivity with the problem of determinism [7]: a realist can consistently maintain an indeterministic stand to any extent, just as a subjectivist can be as much of a determinist as he pleases. For realism the precise behavior of physical objects is irrelevant as long as they can walk alone.)

In order to restore realism in physics all we need is to reinterpret the present formalisms of the quantum theory abiding by the rules of semantical consistency and keeping in mind the goal of producing a physical rather than a psychological theory of the microphysical world. This is possible now, without modifying the present formalisms—which need repairs for different purposes. Such a realistic interpretation of the available frameworks is unlikely to take us back to pre-quantum physics. For example, it will not be able to claim that an electron *has* at the same time both a precise position and a precise momentum, only Heisenberg's relations tell us that we cannot measure them, i.e. *know* them empirically with complete accuracy. Indeed, if Heisenberg's relations are assumed to hold for every mechanical system, whether under observation or not, a realist cannot, on the strength of the usual theory, assign the electron a simultaneous precise position and momentum. That is, he cannot regard it as a classical point-particle—which we also know from diffraction experiments with extremely weak particle beams.

We are now in a position to handle the problem of physical interpretation in a more complete and precise way than was done in Section 4.

6. Reference and Evidence [8]

We should know both what a physical theory is supposed to represent and what does sustain the claim to such a reference, i.e. what its evidence is. If we concentrate on the reference we may end up in uncritical realism, whereas if we ignore the reference we are bound to wind up in subjectivism.

Referentially (semantically) considered, a physical theory points in an immediate way to a conceptual model which is in turn supposed to symbolize a real system of some kind (see Section 3).

Just as the immediate referent is a construct, so the mediate referent can in fact be nonexistent, and in any case it need not be observable. And *evidentially* (methodologically) considered, the same theory points in a devious way to a set of observed and potentially observable facts—the available and the possible empirical evidence (see Figure 2). It is not just that a physical theory should say *more* than whatever is expressed by the set of actual empirical information that triggers and tests the theory since otherwise it would be just a summary of information on the latter's level: a physical theory is supposed to say things quite *different* from the observational reports relevant (favorably or unfavorably) to it. Thus, atomic theories are not about spectroscopic observations although they participate (alongside other theories) in the explanation of such data.

For example, the immediate referent of the kinetic theory of gases is any member of a certain set of idealized ensembles of particles assumed to possess certain characteristics, whereas one of the mediate referents of that theory is a nebula. Data concerning nebular kinematics constitute part of the evidence for or against the kinetic theory and/or the hypothesis that it applies approximately to such

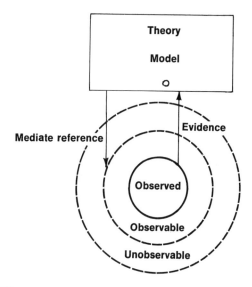

Figure 2   Reference and evidence different.

systems—a metatheoretical assumption, by the way. Any future report of the same kind will be a further evidence of the same kind.
In this case the intended referent of the theory can be observed
with the help of instruments built and interpreted with the help
of further theories, chiefly mechanics and optics, which play here
an instrumental rather than an explanatory or substantive role.
Another example: any theory of "elementary particles" refers mediately to certain unobservables suspected (hypothesized) to be entities (real existents), but it only supplies a hypothetical and most
likely a rather crude model of them. And the evidence relevant to
such a theory—e.g., a set of tracks on a nuclear photographic plate
—differs in nature from the referents of the theory: the tracks are
not referred to at all by the theory and these data *become* an
evidence relevant to the theory on condition that they be interpreted
in the light of another body of theory (notably classical mechanics
and some theory concerning the passage of electrically charged particles through matter).

In this respect the job of the physicist is not unlike the task of
the paleontologist, the historian, or even the detective: in all these

cases unseen facts are hypothesized and such hypotheses and systems of hypotheses are tested through the observable traces left by the presumptive criminal (extinct animal, hero, or proton), which traces become evidences only in the light of instrumental or auxiliary hypothesis and/or theories concerning the possible mechanisms whereby the traces might have been produced; clearly, the theory under test may take place in such an explanation, i.e. it may contribute to produce its own evidence.

(So far we have purposefully avoided talking of *phenomena* as data, much less as evidence relevant to physical theories. The reason is this. What philosophers call a *phenomenon* is an event occurring in connection with some cognitive subject: phenomena are what appear to us, humans, whence no humans, no phenomena. *Phenomenalism* is the doctrine according to which the world is the set of appearances; in particular, physical reality would be the set of observations conducted by physicists. The programme of phenomenalism, shared to a large extent by operationalism, is the construction of physical objects as systems of appearances. This programme has failed and is unfeasible. There are various reasons for rejecting phenomenalism, among others the following. Firstly, physics is not interested in what appears to me, or in what seems to me to be the case: physics is an attempt to transcend subjectivity, to go beyond perspectivism. Secondly, most phenomena involve macroscopic events which in principle can be explained jointly by physics and psychology. Thirdly, the programme of phenomenalism has failed, whereas the programme of realism, of explaining appearance by (hypothesized) reality is working. Fourthly, phenomena are lawless; only objective (largely unperceivable) facts are supposed to be lawful, and there is no science which is not a pack of law statements.

To return to the differences between the hypothetical referent and the observational evidence of a physical theory. For one thing, the intended mediate referent of a theory is supposed to exist independently of the theory—which assumption may be false. On the other hand there can be no evidence without some theory or other, however sketchey, since the theory itself will determine whether or not a given datum is relevant to it. (What theories playing an instrumental role do is to help gather and interpret such data, but the relevance of the data to the theory under test is determined by the latter.) Thus, a quantum theory of "particle" scattering will be a ground for accepting measured values of directional beams, and if the projectiles are assumed to be electrically charged, the measurable curvature of the visible tracks they leave behind will also be considered relevant—just because the theory, in conjunction with classical electrodynamics, says so. On the other hand thousand other pieces of information regarding the same experimental set-up will be quite irrelevant to the theory under test—which is a blessing. Observation reports (data) must be interpreted with the help of at least one theory in order to become evidence. If preferred, the empirical support a given substantive theory enjoys is determined by comparing the latter's predictions with the evidence produced by empirical operations designed and interpreted with the help of at least one theory (see Figure 3). Another way of putting it is

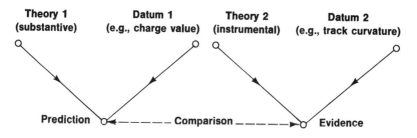

| Theory 1 (substantive) | Datum 1 (e.g., charge value) | Theory 2 (instrumental) | Datum 2 (e.g., track curvature) |

Prediction — — — — — Comparison — — — — → Evidence

Figure 3   Theories (substantive and instrumental), data, and evidence.

this. No single fundamental theory can directly explain observations, i.e. can do it without the assistance of further theories. (What a single theory can explain are *thought experiments*, such as the two-slit "experiment," the actual performance of which requires, in the case of "particles," the employment of real crystals and consequently some theory about crystal structure and another theory about the particle-screen interaction mechanism producing the observable fringes.)

To sum up: 1) fundamental physical theories have no observational content, i.e. they contain no purely observation statements, and consequently they cannot be reduced to sets of data or even to data-processing devices; 2) there is no theory-free evidence in physics. If this much is accepted, we need not confuse reference with evidence.

7. Rules of Interpretation [8]

If the preceding analysis is accepted, then it must be acknowledged that in physics we meet more than one kind of rule of interpretation (semantical rule). The mathematical formulas of physics can be read with the assistance of interpretation rules of two kinds: referential and evidential. A *referential interpretation rule* (*RIR*) establishes a correspondence between some of the theory's nonformal symbols, and its referent. Consequently, a rule of this kind contributes to the (core) meaning of the theory; in the ideal case of a very simple theory, the set of its referential interpretation rules makes up the full physical meaning of the theory. On the other hand an *evidential interpretation rule (EIR)* links a low level theoretical term to some observable entity or trait, such as a visible clock.

Since every physical theory has both an immediate and a mediate or intended referent (see Section 3), we must distinguish two types of referential interpretation rules: 1) *type* I *RIR*'s establishing correspondences between nonformal concepts and traits of the ideal model (which are further concepts rather than real things or properties), and 2) *type* II *RIR*'s, establishing correspondences between traits of the theoretical model and features of the latter's hypothesized real referent. Example of *RIR*'s: the geometrical concept of reference frame is interpreted as a rigid trihedron (type I *RIR*), which ideal object is in turn interpreted as an approximate model of a real semirigid body (type II *RIR*). Example of an *EIR:* a peak in an oscilloscope graph is sometimes interpreted as an effect of an electric discharge (see Figure 4).

The referential interpretation rules are necessary, though insufficient, to delineate the meaning of a theory: they indicate what may be called the *core meaning* of the symbolic system. (A full determination of its meaning would require unearthing all the presuppositions of the theory as well as actually deducing the infinitely many consequences of its initial assumptions—none of which operations is effectively possible [9].) In order to determine the *testa-*

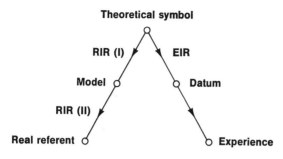

Figure 4   Physical interpretation rules.

bility of the theory, and a fortiori in order to carry out actual empirical tests of it, we must add a set of evidential interpretation rules. But usually such *EIR*'s are not among the formulas of the theory, because every evidence depends not only on the given theory but also on other theories (see Section 6) and on the available hardware. Thus precision length measurements may require interferometers and electronic circuitry as well as several fragments of theory built into them and enabling us to operate and read them.

(The fundamental quantum theory includes *RIR*'s but it does not and should not contain *EIR*'s: it is not capable of being interpreted operationally since it applies chiefly to unobservable physical objects: recall Section 5. On the other hand some of its applications, e.g. the half-born quantum theory of measurement, and solid state theory, do contain some *EIR*'s, as such theories constitute chains between a hypothesized microlevel and a partially observable macrolevel.)

If the preceding distinctions are ignored, the complex character of the relation of physics to reality, or else its roundabout relation to experience, can be overlooked. The intended objective reference is insufficient to ensure testability, let alone truth. Thus, any number of "elementary particle" theories have been proposed, and will continue to be advanced, which are either too far from currently possible tests or collide head-on with available data. Unfortunately there are no foolproof *a priori* warrants of objectivity, such as invariance under certain transformations; even the invariance of the basic equations with respect to observer changes is necessary but insufficient to attain objectivity [5] The adequacy of the objective reference of a theory must be estimated with the help of experience and further theories: those logically presupposed by the given theory and those employed in its empirical test [10].

By focusing on the test of a theory we do not eliminate the question of its objective (yet hypothetical) reference. In fact, testing presupposes the objective reality of at least the instruments handled by the operator—an assumption which the theory itself may not make, unconcerned as it may be with any empirical operation. Thus, a theory about the field generated by a one-dimensional antenna deals with an immediate referent which is a grossly simplified model of a real rod-like antenna. We do not assume the physical existence of the model described by the theory, but the test of the theory will call for the manufacturing and operation of a number of supposedly real instruments, no less than of a real rod-like antenna —the intended referent of the theory. A test involving neither the intended real referent nor actual pieces of apparatus is not an actual empirical test but a thought experiment (e.g., a simulation on a computer). In short, scientific experience presupposes the reality of the objects it manipulates, even if it does not commit itself to the hypothesis that a given theory's intended referent is actual: after all, the test aims at testing such a hypothesis. Since we are at it we may take a further step and claim that experience is a proper, and even a smallish, subset of the total set of facts, and that physics handles some of them—as well as wholly imaginary ones (see Figure 5). This may be regarded as the Kernel of critical realism.

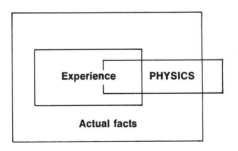

Figure 5   Physics covers some facts, among them observable facts, and it stretches over nonexistent facts (falsely assumed facts).

8. Concluding Remarks

Every theory is about objects of some kind—which it identifies as members of a universe of discourse U—to which it assigns certain definite and basic properties P_i—are largely or wholly nonobservational, i.e. they fail to have experiential counterparts—as exemplified by "mass point," "mass," "electric charge," and "isotopic spin." U and the P_i are the building blocks of initial assumptions (postulates) that refer to U itself, the immediate referent of the theory. From such initial hypotheses, in conjunction with auxiliary premises (such as data), logical consequences are drawn with the help of the underlying logical and/or mathematical theories. Up to this point, a well-organized (semiaxiomatized) physical theory does not differ from a mathematical theory.

The key differences between a mathematical and a physical theory are two: one is semantical, the other methodological. The former consists in that, although neither the U nor the P_i of a physical theory are snapshots of physical objects, they purport to symbolize them: the members of U are expected to mimic real objects, and the theory as a whole is expected to represent the behavior and/or structure [11] of these ultimate targets of physical theorizing. In short, physical theories, unlike formal ones, are expected to have a real (yet mediate) referent in addition to a conceptual (immediate) referent.

The methodological difference consists in that some of the logical consequences of the initial assumptions of a physical theory should be susceptible to empirical test. Every such test involves not only the direct control of certain observable features but also 1) the frequently indirect control of the intended (mediate) referent of the theory, and 2) existence hypotheses concerning one or more links in the chain that goes down to the hypothetical real referents of the theory. An empirical evidence relevant to a theory can differ from the latter's intended referent as much as a clinical syndrome can differ from the corresponding illness. The outcomes of empirical tests, alongside theoretical considerations (e.g., compatibility with contiguous accepted theories) and metascientific ones (e.g., consistency with the prevailing philosophical tenets) provide some evidence enabling us to argue about the degree of truth of a theory, i.e., the extent to which its mediate reference is adequate [21].

The intended objective reference and the empirical test of a physical theory are distinct but they come together: no real referent (however hypothetical), no point in either theorizing or testing; no test, no possibility of estimating the degree of truth of the real referent hypothesis. This adequacy cannot be expected to be complete if only because a physical theory is built by inventing a simplified and thoroughly hypothetical model of the intended referent. (Even a black box is such a model.) The realization of this unavoidable imperfection motivates the invention of richer, usually more complex theories, some of which succeed in getting closer to the objective referent, while others miss the target even worse [13].

In short, every physical theory 1) is built with symbolic (noniconic) and partly conventional units; 2) it is supposed (often wrongly) to refer ultimately to real objects (physical systems; 3) its reference to such real objects is incomplete, extremely indirect, and at best partially true, and 4) its test involves further theories and presupposes the physical existence of certain objects. Any hypothetico-deductive system has the first of the preceding properties, but if it fails to have all four of them then it is not a physical theory.

These traits of every physical theory—and, indeed, of every factual theory—render uncritical realism obsolete, but they also make subjectivism (idealism, conventionalism, fictionalism, phenomenalism, operationalism, etc.) untenable. These latter, nonphysical, views of physics were made possible by the failure of naive realists to recognize the symbolic character of physical theory, the hypothetical, indirect, incomplete and global (rather than isomorphic) character of its reference to physical systems, the partial adequacy of such a reference, and the physical basis of the empirical operations whereby such adequacy claims are put to the test. Once these traits of physics are recognized, the accounts of it advanced by uncritical realism and by the various shades of subjectivism are left behind as so many partial views—simplistic views—of our science. Note that these views are not here discarded on the strength of philosophical tenets and arguments of a traditional kind but with the help of mathematical logic, semantics and methodology, the very tools which were once thought to support the subjectivistic philosophies of physics.

The vacuum left by the (logical) passing away of the above-mentioned philosophies of physics should be filled by building a theory of knowledge subsuming and expanding the seeds of truth contained in the former doctrines, as well as the realistic hypotheses presupposed and suggested by science [14, 15]. Such a critical or scientific realism could assist (and in turn be tested by) the construction of a thoroughly physical (rather than psychological) interpretation of the mathematical formalisms of the quantum theory. The new realistic epistemologist should both stick his neck out and be prepared to have it cut: he should advance daring (yet grounded) assumptions, none of which he should regard as uncontroversial. Thus, while the naive realist may take the reality of electrons for granted, the critical realist will say that *present*-day physics *assumes* that electrons are real things—i.e. he hypothesizes that the electron concept has a concrete counterpart but, at the same time, he would not be too surprised if this assumption turned out to be false and electrons were replaced by something else.

Needless to say, whatever form the new critical realist epistemology may take, it would fail to meet the standards of scientific research and consequently it would fail to help this enterprise if it were conceived as one more *ism*, i.e. as a set of tenets beyond criticism and above science. Wanted: a name for this nascent epistemology, one not ending in *ism*, for whatever ends in *ism* is apt to put an end to the quest for truth.

REFERENCES

[1] Cf. the author's *Scientific Research* (forthcoming in Springer-Verlag, 2 volumes)—quoted in the following as *SR*—Secs. 2.2, 2.3, 3.5 and 3.6.
[2] Cf. *SR*, Secs. 7.1, 8.1 and 8.4.
[3] Cf. the author's *The Myth of Simplicity* (Prentice-Hall, Englewood Cliffs, N. J., 1963), Part II.
[4] Cf. *SR*, Secs. 3.5, 3.6, 3.7 and 7.5.
[5] Cf. the author's *Metascientific Queries* (Charles C Thomas, Springfield, Ill., 1959), Chap. 8.
[6] Cf. *SR*, Sec. 7.2.

[7] Cf. the author's *Causality: The Place of the Causal Principle in Modern Science,* 2nd ed. (Meridian Books, Cleveland and New York, 1963), Appendix.

[8] Cf. *SR,* Sec. 8.4.

[9] Cf. the author's *Intuition and Science* (Prentice-Hall, Englewood Cliffs, N. J., 1962), pp. 72 ff.

[10] Cf. *SR,* Secs. 8.4, 12.4, 15.6 and 15.7.

[11] For the phenomenological and the representational styles of theorizing, cf. *SR,* 5.4 and 8.5, and *Phenomenological Theories,* in *The Critical Approach to Science and Philosophy,* ed. by M. Bunge in honor of Karl R. Popper (Free Press, New York; Collier-Macmillan, London, 1964).

[12] Cf. *SR,* Secs. 15.6 and 15.7.

[13] Cf. *SR,* Secs. 8.1, 8.4 and 8.5.

[14] Cf. *SR,* Sec. 5.9.

[15] See the following recent defenses of critical realism: P. Bernays' contribution to the discussion on quantum mechanics in the *Revue de Métaphysique et de Morale,* Avril–Juin 1962; H. Feigl, *Matter Still Largely Material, Philosophy of Science* 29, 39 (1962); K. R. Popper, *Conjectures and Refutations* (Routledge and Kegan Paul, London; Basic Books, New York, 1963); and J. J. C. Smart, *Philosophy and Scientific Realism* (Routledge and Kegan Paul, London, 1963).

Patrick A. Heelan

Patrick A. Heelan was born in Ireland and joined the Jesuit order as a young man. After receiving a Ph.D. in physics from St. Louis University and a further degree in theology, he spent two years in the United States as a Fulbright fellow doing post-doctoral research in theoretical physics. He received a Ph.D. degree in philosophy from Louvain University in Belgium and is currently chairman of the Philosophy Department at the State University of New York at Stony Brook. The present selection, two chapters from his study of the philosophy of Werner Heisenberg, represents a critical realism in the tradition of neo-Thomism established by Maréchal, Bernard Lonergan, and others. On the basis of this philosophical position Heelan evaluates the philosophy of Heisenberg and presents his own interpretation of the nature of scientific knowledge and the reality it reveals.

Quantum Mechanics and Objectivity

ONTOLOGICAL STRUCTURE OF PHYSICAL REALITY

Section I: Justification of Realism in Physical Science

'Reality' and Human Polymorphic Consciousness

If there is one conclusion which imposes itself before all others as a result of the inquiry we have made, it is the ambiguity hidden in the sense of the term 'physical reality'. This is founded in turn upon the underlying polymorphism of the human way of knowing reality. The neglect of some elements of this polymorphic consciousness, or undue emphasis on certain aspects of it are the roots from which spring a multiplicity of epistemological difficulties; for in every question there is a hidden structure directing implicitly the search for answers, and prior even to the formulation of the answer and imposing a structure upon the answer even before it is formulated. This hidden structure is the domain of intentionality and, like the nine-tenths of an iceberg below water, it lies perilously below the level of our cognitive activities. Because of failure to appreciate this, many ontological and epistemological discussions, especially wherever positivism or linguistic analysis is dominant, tend to foun-

From chaps. 9 and 10 of Patrick A. Heelan, S.J., *Quantum Mechanics and Objectivity: A Study of the Physical Philosophy of Werner Heisenberg*, pp. 156–79, by permission of Martinus Nijhoff.

der; for, in such cases, problems are generally formulated uniquely in the light of the one-tenth that is in public view.

By saying that the human way of knowing is polymorphic, we mean that it is a composite act in which many different activities take part in subordinated roles. Three activities are particularly prominent in such an act; viz., empirical intuition, conceptual understanding and rational affirmation (or judgement). Still it is one act and not many acts; since its purpose is to express a single, if composite, object of knowledge. However, the polymorphism of human consciousness creates many problems. Some of its elements, for instance, may be overlooked. Or alternatively, the diversity of its activities may be recognised but divided into two classes: one constituting the class which really knows *reality* and which can be used to define the sense of the term 'reality' (since reality is, by supposition, its object), and a second class which is considered inessential to this knowledge either because it constitutes merely a provisional or imperfect stage of knowledge or because it provides us merely with criteria of reality which are subjective and extrinsic to the essential definition of reality. We have had occasion to describe the two extremes of simplification: *empiricism* which tends to devaluate conceptual knowledge and to emphasise the act of empirical intuition, as the essential cognitive act defining (for us) the sense of the term 'reality', and *rationalism* which tends to devaluate empirical knowledge and to see the mind's essential function as that of defining conceptually the essences of real things.

We admit both types of knowledge as essential to the expression of concrete physical reality and by so doing place ourselves within the perspective of the *abstractionist theory of knowledge* outlined in Chapter One. That is, we hold that conceptual understanding is not itself alone an adequate expression of concrete physical reality; but that its function is to yield an ideal relational norm compared with which individual cases depart only randomly. The function of recognising individual cases is performed by empirical intuition. It follows from this, as we have shown, that the most complete knowledge we can have of a physical system is the combination of a deterministic theory (specifying the ideal norm) and a statistical theory (specifying the ideal frequencies of departures from the norm within a random sample of individual cases).

Heisenberg's Ontology Criticised

Heisenberg, after an initial phase of empiricism, rediscovered the ontological value of rational thought. The discovery was made however under the influence of Plato from whom the dual objects of knowledge, objects of sensibility and objects of understanding were strongly contrasted and opposed. The influence of Kant caused him to admit yet a third class of possible objects, namely, noumenal objects. So much cognitional wealth was a mixed blessing for it created the problem of harmonising the diverse elements and of stating their relation to one another and their ontological value. The problem for Heisenberg was rendered practically insoluble by reason of an implicit supposition that the three objects of knowledge were three fully constituted, if linked, objects, and not three aspects entering into the constitution of one object. Within the Kantian problematic as Heisenberg understood it, 'reality', as the aim and objective of our knowing powers, tended to have a triple degeneracy which could be resolved in different ways: by the simple process of suppressing one or other element in the degeneracy, or by the reversal of all three in favour of the exclusive rights of the self-conscious subject to be called real.

'Reality': a Complex Notion

Our own solution is to retain all the elements of the problem but not as a trio of fully constituted objects; rather as revealing complementary aspects of the strict object. The sense in which we use the term *strict object* has already been explained.[1] We shall now review the different functions performed by the three partial aspects (or partial objects) of human knowledge within the whole.

Sensibility

The empirical data received in sensibility serves the twofold function of making present the existence of the reality to be affirmed,

1. Patrick Heelan, S.J., *Quantum Mechanics and Objectivity: A Study of the Physical Philosophy of Werner Heisenberg* (The Hague: M. Nijhoff, 1965), chap. i, pp. 6–8.

recognised or understood (its *presentative* function), and of being a symbol (or part of one) of the nature of the reality presented in and through it (its *representative* function).

In its *presentative function,* sensibility is related to the judgement of existence and is connected with the criterion for us of the affirmation of existence. For example, we can only affirm that a hydrogen atom is in this piece of material if we have some sensible and unambiguous indication of its presence, i.e., an observable symbol of its presence. In the case of the hydrogen atom, this may be a certain spectrum, a certain atomic number, the ability to take part in certain chemical reactions, etc.

In its *representative function,* empirical intuition forms an essential part in the constitution of the observable symbol. The object, in the strict sense, with which we have identified the sense of the term 'reality' is that which is symbolised by the observable symbol. It follows from this that there is an aspect of physical reality to which sensibility corresponds and that this aspect is not extrinsic to the ontological constitution of physical reality. Observed data in physics serve to represent individual systems: therefore, the representative function of sensibility is to represent the individuality of concrete cases.[2]

Observed data, however, always exhibit an unsystematic aspect with respect to the general law (or ideal norm) to which they more or less conform. The element of randomness in observed data with respect to a general law must also have its representative value. The element of randomness is not merely an expression of human ignorance but an expression of the essential and characteristic structure of human knowing. We argue that there must be an ontological foundation in physical reality which enables it to be known and truly represented by a strict object possessing the structure in question. The ontological foundation must be composite, with a formal or systematic principle or *form* to explain the systematic normative structure of the physical law, and a principle which we call *irreducible matter* to explain the openness and multiplicity of the kind of physical system in question.

2. The essential reference to sensibility introduces into the meaning of 'reality' the *temporality* which many modern philosophers give as the meaning of 'existence'. Cf. A. Dondeyne, *La différence ontologique chez M. Heidegger,* p. 24.

Conceptual Understanding

The rational conceptual part serves to add to the data presented in empirical intuition (or to recognise in them) a structure of relations and terms (i.e., a system of *primary relativities*) and an *ideal norm* to which the actual structures conform more or less (this is the *secondary determination of the above*). Consider, for example, Galileo's experiments on projectiles.[3] He surmised that a parabolic relation existed between the vertical distances traversed and the time of flight: the primary structure here is one constituted by a mathematical functional with parameters to be determined. The secondary determination in this case is the explicit value of the constant parameters. The secondary determinations can be crude and inexact and subject to indefinite improvement by improved techniques of measurement. This does not falsify, however, the primary relativities.

The relational structure in question may be of two kinds. It may involve relations between things and things, or it may involve relations between things and certain aspects of sensible experience or utility to a human subject. In both cases, the effect of conceptual analysis is, firstly, to express or to suppose a set of terms and the primary relativities between them, and, secondly, to make precise the ideal normative inter-relation of this set with a view to achieving clarity and precision, e.g., by evaluating the secondary determinations. In physics, precision is obtained by measuring more and more accurately the unknown parameters of the equation.

We distinguish two classes of such structures: one which constitutes a thing-for-us or property-for-us, and the other which constitutes a thing-for-things or property-for-things.[4] In the first case, since the content of a concept is precisely to relate things or properties to a human subject, it may well be thought that such a concept does not express what a thing is, but merely how it appears to me or how usefully I can employ it. So speaks the instrumentalist philosopher. The reasons he gives would hold if the relations through which we expressed our knowledge of physical reality were

3. Galileo Galilei, *Dialogues Concerning Two New Sciences* trans. by H. Crew (London: Macmillan, 1914), the "Fourth Day."
4. Heelan, chap. iv, pp. 58–68.

merely extrintic relations, that is, relations founded upon a ground extrinsic to the nature of the related term; as for example, when we speak of a distant object, *distance* is extrinsic to the object considered. However, conceptual knowledge usually prescinds from extrinsic relations in order to construct concepts based upon the activities of things whether to form impressions on human sensibility, to fulfil human desires or to effect changes in the world. Such relations are founded upon what things do, and hence upon what things are: they are *intrinsic relations*. They reveal, in consequence, something of what things are. In the class of things-for-us or properties-for-us, however, the concept is specified by the value (in experience or in action) of the thing or property for us. The thing or property is known only as the term of a select set of relations, whether to the structure of our senses or to the actual human needs and desires to which things respond and which are characteristic of a certain historical situation.

To complement the selectivity of thing-for-us or property-for-us, natural science enters to study the relationships between things themselves. Essential to this study is the use of symbolic (mathematical) formalisms to express closed groups of relations defining one another mutually by implicit definition. Essential to this study also is the use of observable symbols in experience; both as criteria for the judgement of existence and as tokens of the formal effect of particular interactions. In the latter sense, the token can be mapped on the number-system and thereby incorporated into the mathematical formalism which with the rules of correspondence defines the properties-for-things. Since the relations so defined are all based upon interactivity between things these relations are all intrinsic to the structure of physical reality and reveal something about its essence. In general relativity even space and time become intrinsic properties of physical systems.[5]

We know physical reality, then, to the extent of its participation in a world of actually interacting things and subjects. We know it, however, only through sensible and intelligible symbols. Does our knowledge go beyond the symbol to reach out even to that which is symbolised? Or does it stop short, as in Heisenberg's explanations, at the symbol leaving the thing-in-itself shrouded in mystery?

5. *Ibid.*, chap. x, pp. 172–74.

Rational Affirmation

This brings us to the sense of the third activity constitutive of the strict object of human knowledge, viz., the act of affirmation (or judgement). What is experienced and what is conceived are only partial and preliminary activities leading—after the weighing of appropriate evidence—to the culmination of a virtually unconditioned affirmation "This is so." [6] *What is so,* however, is not just a phenomenal object nor a symbol in its material reality, but what the phenomenal object *represents* and what the symbol *symbolises.* Thus, the mind turns about the pivot of the observable symbol and the organization of such symbols through a mathematical theory, to affirm that which so presents itself in experience, viz., the thing as transcendent to experience and thought and existing independently of whether it is experienced or thought about.[7] This follows from the fact that the concepts of physical science are not things-for-us or properties-for-us, but relationships between things; hence the sense of the affirmation "This is so" expresses what is, whether or not it is experienced or thought.

One might object that even though this be the sense of the affirmation, it could be mistaken—even mistaken as a rule. How do we know that the affirmation does not make its own strict object by setting off its own conception on its own authority against a horizon of absoluteness which is no more than the projection of an ideal and immanent regulative principle governing all human affirmations? While we do not wish to enter here into a lengthy discussion of the critical problem, a few brief points might help to show where we, at least, stand.

First of all, a universal scepticism about the realistic value of

6. In other words we identify the meaning of 'physical reality' with that of 'existing thing'. However, *existence* is not a mere predicate: its sense is expressed by the strong sense of the verb *to be.* We are are critical however of some of the implications of G. Maxwell's formula: "Φ_s are real $=_{def} \Phi_s$ exist." This seems to us to imply that the valid application of the term *existence* is limited to *things,* thereby excluding *spirit* as a possible subject of existence. If *existence* in the above formula is restricted to *physical existence* then we should be in entire agreement with the formula. The reference is to "The Ontological Status of Theoretical Entities" by Grover Maxwell, *Minnesota Studies in the Philosophy of Science,* III, p. 21. Cf. also note 7 below.

7. Cf. Joseph de Finance, *Être et Agir* (Rome: Univ. Greg., 1960), chap. III.

human affirmations is untenable; for, as Augustine saw, to doubt means *really* to doubt and not doubt one's doubting; hence universal scepticism is self-contradictory. If universal scepticism is self-contradictory, then there exist criteria according to which affirmations may be judged to be well and correctly made or badly and incorrectly made. The sense of the affirmation, however, whether it is well made (and hence true) or badly made (and hence false) remains the same, viz., to assert *what is* independently of whether or not it is an object of my thought or experience. The criteria for correctly making affirmations is the weighing of evidence. This evidence is often encountered in haphazard fashion in everyday life. In scientific research, however, the business of the scientist is to produce evidence in a systematic fashion, reconnoitering the terrain of a scientific theory strategically, occupying well chosen and strategic points with a view to undisputed mastery of the terrain. No one knows better what this means in any particular field than the trained scientist. The scientific method of experimental testing and verification is the search for and the application of criteria for the correct making of scientific affirmations; but the sense of the scientific affirmation whose justification is sought through these procedures is incurably *realistic,* that is, it intends being as its object.[8]

In conclusion, it may be said that if any scientific affirmation is well and correctly made, then it asserts *what truly is.* It may, however take a great deal of difficult and subtle analysis to disengage the physiognomy of the strict object which is affirmed from the matrix of scientific methodology within which it makes its appearance. In our opinion, the real source of epistemological difficulties in the interpretation of quantum mechanics lies here, rather than in disputes about the sense of the act of affirmation or judgement in human knowledge.

Section II: Ontological Structure of Atomic Reality

Whole and Part of an Atomic System

In what sense are the constitutive parts of a bound many-particle system present in the system? Or in the concrete, in what sense are the proton and the electron present in the hydrogen atom? Is the

8. Cf. B. Lonergan, *Insight,* chapters XI–XIII where the questions discussed in this section are very fully treated.

hydrogen atom merely a compound, perhaps a mechanistic compound, of a proton and an electron? Heisenberg's opinion which we have already reviewed is that there is no essential difference between an elementary and a compound atomic system.[9] That which imposes unity on a system, he says, is a mathematical form and the mathematical form of a hydrogen atom is not just the juxtaposition of two forms, but the excitation of a two-particle (proton and electron) bound state in a fundamental mater (or energy) field. We agree with Heisenberg's conclusion, although we do not concur with the immanentist interpretation he has given to it.

We can reach a similar conclusion by another route. If the proton were an actual physical reality within the atom, then, it should be able to manifest its presence through characteristic types of interactions with neighbouring systems. The neighbouring systems would thereby serve the function of a measuring apparatus: the characteristic effects of the proton are its formal effects on a neighbouring system. These can be magnified and transformed by suitable physical instruments into a symbol which can be observed by us. Quantum mechanics, however, tells us that if a proton forming part of a hydrogen atom is to interact with its environment in the way characteristic of a proton, the hydrogen atom must absorb an energetic impulse which destroys it as a bound system. Hence, in the moment in which the proton manifests its independent presence in its physical milieu, the hydrogen atom ceases to exist. The observable symbols of a hydrogen atom and of its proton-nucleus cannot be simultaneously produced. The significance attached to this is that either a hydrogen atom exists or a proton and an electron exist. What is described as a bound state of two particles does not leave the particles intrinsically unaffected by the union. It is the production of a new physical system, unique and distinct from its component parts. Its form is properly neither elementary nor compound as Heisenberg has pointed out; for these terms connote a materialist or reductionist explanation which simply does not apply to the compound system. The explanation given in this paragraph of the relation of part to whole in a compound corresponds more or less to the philosophic notion of *virtual presence* as, for example used by Aquinas.[10]

9. Heelan, chap. viii, p. 147.
10. Cf. Aimé Forest, *La Structure métaphysique du concret selon Saint Thomas d'Aquin* (Paris: Vrin, 1956), p. 199.

Energy: A "Universal Matter"?

We have briefly discussed the formal principle of a physical system: what then is its *irreducible material principle?* As energy is the only common non-vanishing factor of every physical transmutation, Heisenberg postulated that it plays the part of a "primary matter" or "universal substance." This acts like a basic field capable of being informed by any one of a series of stationary states (particle states) described by mathematical operators. This view is also put forward tentatively by Lonergan, who supports his view by pointing out that energy is a physical invariant constructed by integration but integration is a mathematical operation which tends to abolish formal structure while differentiation on the other hand complicates the formal structure.[11] For this reason, the total energy is independent of whether it is realised concretely as an atomic system or, for example, as radiation or kinetic energy of motion. The total energy is rather the limit which specifies what variety of forms is possible. In elementary particle physics, it is generally supposed that the system spends a proportion of its time in each of its possible states (or "channels") much as a canonical ensemble in thermodynamics is thought to take on every possible energy distribution. Lonergan suggests that energy may be the name given to (Aristotelian) *primary matter in the concrete,* i.e., the name for the concrete limitations of a physical system imposed by its primary matter.

While acknowledging the universal role played by energy in physical processes, we do not, for the following reasons, think that its role is analogous to that of Aristotelian and later scholastic primary matter. In the first place, energy is not the only universal determinant of physical transformations. Baryon (or atomic) number, lepton number, electric charge are also universal invariants, with this difference, however, that they may vanish for a particular system. However, there seems to be no essential difference between energy which is the invariant corresponding to Lorentz symmetry, and say, charge, which is the invariant corresponding to a certain

11. B. Lonergan, *Insight* (London: 1957), pp. 443–444. The opposite view, viz., that energy is a *universal act,* is held by L. B. Guèrard des Lauriers, in "La substance sensible," *Angelicum* xxxix (1962), pp. 40–91, 350–394.

kind of gauge symmetry. In the second place, energy is not a simple quantity but a component of a four-vector, indicating its essential relativity to a frame of reference. The frame of reference is provided by a measuring instrument representing the physical milieu of the system in question. Energy seems rather to be a *condition of possibility* specified by the milieu and limiting what can take place in the milieu.

The "Energy Ladder"

The operation of this condition of possibility in different milieus is illutrated well by Weisskopf's "energy ladder." [12] The free energy capable of being exchanged between the physical system and its milieu sets limits to the character of the systems capable of existing in a stable fashion in the milieu. For example, starting at the lowest "rung" of the "energy ladder," the existence of complicated heavy molecules, like proteins and amino-acids, depends on the fact that neighbouring systems cannot exchange more than approximately 0.01 e.v. (electron volt) of energy. In this physical environment the heavy molecule reacts as a stable whole, with characteristic properties vis-à-vis the class of its interactions with its environment. Its parts are virtual, not actual, parts. The representation of it then as a structure made up of parts is from the epistemological point of view merely a symbolic representation.

If the energy capable of being exchanged between the system and its environment is increased from 0.01 e.v. to approximately 0.1 e.v., the heavy molecules become unstable and disintegrate, actualising in the process a set of simpler chemical compounds and molecules. In this state, the re-formation of a heavy molecule is not ruled out, for there is a non-zero probability of emergence of the heavy molecule out of the mixture, although there is a much larger probability of decay. As the free exchange energy is further increased even the chemical compounds break up, forming first a mixture of partly ionised atoms (at around 5 e.v.), and then a plasma of stripped nuclei (at around 10^6 e.v.) and finally, at even higher energies, a plasma of free nucleons, electrons and other elementary particles with particle-antiparticle pair production and annihilation.

12. Victor Weisskopf, "The Quantum Ladder," *International Science and Technology*, June, 1963, pp. 62–70.

The role of energy in the "ladder" is not primarily to provide the "substance" out of which the systems in echelon are made, but rather as a condition of possibility specified by the milieu and limiting the kind of system or process permitted in this milieu.

Irreducible Matter

What then is *irreducible matter?* We arrived at the notion of "irreducible matter" by analysing the conditions of possibility of the strict object of quantum mechanics, viz., of the peculiar combination of deterministic and statistical properties which define a quantum mechanical system.[13] Irreducible matter is a constituent of physical reality, but no systematic effects can be attributed to it. It is the reason for the irreducibly statistical character of the observed events which represent to us the behaviour of individual systems. It is the reason for the openness of any physical form to indefinite multiplicity and the basis for the description of a physical system as a member of an equivalence class or as a virtual ensemble of systems. Beyond this, nothing more can be said. One is reminded of the famous account which Aristotle gave of *primary matter: "By material,"* he wrote, "I mean that which is in itself not a particular thing or a quantity or anything else by which things are defined." [14] Irreducible matter is not definable since it has no systematic consequences. But it is not nothing; its only function is to make possible the virtual ensemble which is presupposed by every quantum mechanical state functions.

Summary

This chapter treats briefly three fundamental problems which have arisen during the course of the preceding chapters. The principles for the solution of these problems are now outlined. First of all, there is the problem of the realism of scientific theories. We defend an abstractionist theory of knowledge but insist at the same time on the unity of the polymorphic cognitive activity in which we know physical reality. The structure of this activity itself reveals its

13. Heelan, chap. v, pp. 109 ff.
14. Met. vii (Zeta), 1029a20. Cf. also S. Mansion, *Introduction à la physique aristotélicienne* (Louvain, 1945), p. 74.

realistic tendency. Secondly, we examine the kind of structure which an atomic system possesses, the relation of its parts to the whole, and the function of energy as a universal invariant. We reject the thesis that *energy* is a *universal material substance* (or *primary matter*), and propose instead that it is an ontological condition of possibility established by a milieu, which governs the kinds of systems and processes capable of taking place within the milieu.

LOGIC AND LANGUAGE OF SCIENCE

Section I: The Nature of a Physical Science

The Mathematicisation of Physics

Looking back at the various problems studied in this book, one conclusion at least is evident, that the epistemological structure of modern physical science is one of extraordinary complexity. Physical science in the form with which we are familiar was understandably a late arrival on the historical scene, so late in fact that the beginning of the scientific movement coincides with what historians call *modern times*.

The secret of the success of modern scientific research is the natural affinity between mathematics and natural science. The Pythagoreans are usually given the credit for this discovery, for they stumbled upon the correlation between numbers (or geometry) and certain sensible properties. They found, for example, that the notes of the scale were related to ratios of the whole numbers.[1] This was, however, no more than a minor clue to that mathematicised science which is man's most powerful tool for the transformation of the environment in which he lives. For want of a decisive and momentous insight, the physical sciences of the Greeks remained fallow for over

1. S. Sambursky, *The Physical World of the Greeks* (London: Routledge and Kegan Paul, 1956), chapter 2.

two thousand years. Mathematical physics was delivered into man's hands, as Professor Butterfield says, not by the accumulation of new observational data but "by transpositions that were taking place inside the minds of the scientists themselves." [2]

The period of gestation for the decisive insight was the late Middle Ages. Fourteenth-century writers like Thomas Bradwardine and the Merton College group at Oxford, John Buridan and Nicholas of Oresme at Paris, Albert of Saxony and others, groped mentally for a union of mathematics and physics, which would explain *in a mathematical way* the motions of bodies. During these centuries, a new intentionality was emerging in the minds of scholars and it was one which was at variance with the established outlook of natural philosophy. For four centuries the anticipations of this new intentionality agitated the universities of Europe before it produced its first definite and unqualified success in the mechanical laws of Galileo. Western man had succeeded at last, in Butterfield's phrase, in putting on a new "thinking cap," and, in so doing, he became other than he was with consequences of breathtaking importance and originality. It was the emergence of a new power in man which, more than the religious and political differences, was to shatter the closely woven fabric of the medieval mind. What was this new "thinking cap"?

The Pythagoreans discovered that the musical qualities of sound produced by a plucked string could be mapped on to a set of whole number ratios. Since musical quality is relative to man's hearing, the whole number ratios could be referred in some obscure way to man's ability to recognise harmonies of musical sounds. It was an obscure theory for a number of reasons. First of all, it was restricted to simple arithmetical functions, for it was not clear what a complicated function of whole number ratios would sound like when converted into music—even if such a conversion were possible. Secondly, the ability to judge musical sounds differs from person to person; for it is one of those thing-to-us properties which constitute the content of an observational or descriptive concept. Modern physics does not measure observational properties but only explanatory properties, i.e., properties founded upon thing-to-thing relations. It is even doubtful whether it is meaningful to talk about

2. H. Butterfield, *The Origins of Modern Science, 1300–1800* (London: Bell 1951), p. i.

measuring the quality of musical sounds as these are judged by the human ear. Finally, the theory was of very limited application, for it provided no set of numerical correlates for high or low frequency vibrations, which, though inaudible, are evidently of the same physical nature as the audible vibrations and differ from these only in so far as human hearing does not respond to them. It was because of all these shortcomings that Pythagorean science failed to start a scientific revolution.

The momentous insight which made modern physics possible was the discovery of a set of correlated physical properties, each founded upon a different kind of interactivity between things, which, when mapped upon the number field through suitable measuring instruments constituted a *mathematical functional*. This momentous discovery was to create a new kind of knowledge for which the centre and pole of reference was shifted out of the human observer and into things themselves, principally into macroscopic apparatus called measuring instruments whcih thereby became *observer-instruments*. "The Aristotelians," as Lonergan wrote, "were content to talk about the nature of light, the nature of heat, the nature of weight, etc. . . . where the nature antecedently is specified by a classification based upon sensible similarity . . . Galileo inaugurated modern science by insisting that the nature of weight was not enough; from sensible similarity, which resides in the relation of things to our senses, one must proceed to relations that hold directly between things themselves." [3]

This procedure added three insights of incalculable importance to the Pythagorean discovery. The first was that there existed interrelated groups of physical interactions. The *second* was that the proper effects of these activities in nature, which, when "translated" by circuits, meters, etc. into observable signs, could be mapped on to a symbolic field, like the number field, through the choice of a measuring process and a measuring instrument. The *third* was that the physical interrelation could as a consequence be symbolised numerically (or at least mathematically) with the numerical (or mathematical) symbol placed in correspondence with the observable sign as numbered (or mathematically symbolised). As Professor Ladrière says of the function of mathematics: "Elle ne fournit

3. B. Lonergan *Insight* (London, 1957) p. 38.

pas seulement les procédés de raisonnement admissibles et les cadres des édifices possibles, mais elle fournit aussi les notions de base sur lesquelles ces édifices sont construits. Elle devient ainsi une sorte d'ontologie formelle générale." [4]

Conversely, the existence of a definite mathematical function between measure numbers, argued the existence of a coherent interrelated group of observable symbols and hence of physical properties symbolised.

The structure of human conceptual knowing determines that there shall be two types of physical theories: *deterministic theories* and *statistical theories.* It was the discovery of the quantum theory which brought the realisation that the two kinds of theories might be combined in one formalism.

For the Pythagoreans, the mathematicised musical quality was an observational property, a property-for-us. A physical property in modern science, however, is *defined* within the systematic totality of correlated physical properties with which and in relation to which it is defined. This is at its *explanatory definition.* Each property has in addition an *operational description,* which describes how an appropriate experimental measuring process might be performed, and *observable criteria,* which manifest its presence in the real world, and also provide the (usually numerical) symbol—e.g., a pointer reading—through which it is included in a mathematical theory. It should be noted that only the *explanatory definition* defines *in the strict sense* what the property is: the *operational description* and the *observable criteria* are extrinsic to the property defined. *Observable criteria* merely constitute part of the symbol through which the property manifests its presence in the real world.

Since the scientist's presence in the world is through his body which is a thing in the world, he may sometimes obtain direct intuitive experience through his sensibility of some types of physical properties; as, for example, of light, force, etc. Such intuitive experience can found an observational concept of the property based upon the typical effect of the property on human sensibility or on the typical use it might have for man. This cannot, however, be identified with either the explanatory definition or the operational description or the observable criteria of the physical property. The explanatory definition and the operational description are clearly

4. Jean Ladrière "La philosophie des mathématiques," *Rev. Phil. Louv* LVII (1959), p. 617.

different from this new concept. The observable criteria, moreover, are also different from it, since they describe one of the *symbols of the property in nature*, produced in an instrument and not directly in human sensibility.

Space and Time

The above-mentioned distinction applies to the difference between the physical concept of space and time—in so far as these are explanatory concepts—and the concepts of space and time based upon the structure of human perception and activity in the World-for-man. The first attempt to define explanatorily the structure of space and time was made by Einstein when he proposed the special and general theory of relativity. Up to that time, space and time were conceived to be absolute pre-conditions of physical events and processes. Space was the container of simultaneous events; time ordered sequences of events in an orderly way; but neither space nor time was founded upon an interaction between the events themselves. Absolute Space and Absolute Time were thought to belong to the objective and irreducible elements in any explanation, and probably would have remained this way but for the crisis in physics arising from the non-invariance of Maxwell's electromagnetic equations under the Galilean transformation group which defined the properties of Absolute Space and Absolute Time. Under the influence of Mach, Einstein was led to question the givenness and irreducibility of physical space and time and eventually to analyse the space-time structure presupposed by the scientific method.

Such an attempt, we believe, was bound to be made and was even demanded by the logical structure of the scientific method. The physicist's view of explanation is to construct a World-for-things, viewing reality, as it were, from the point of perspective of an instrumental observer which becomes the centre of its own space and time. Since an instrument is not endowed with an imagination and a memory to situate all things simultaneously and ideally with respect to one another, its "experience" of its physical ambience must be based upon the ways it can actually receive influences from other things and influence other things in turn.

Early in this century it was discovered that the fastest and most universal means of transmitting such influences was, in fact—at least on the macroscopic scale—the electro-magnetic or photon field. Einstein proposed in the special theory of relativity to identify distance with distance-as-measured-by-light-pulses, and time intervals with time intervals-as-measured-by-light-pulses. Since the photon exchange between object and instrument takes a finite interval of time to pass between them, Einstein concluded that the distance between the two things must be related to the time taken by a photon to pass from one to the other. Let *concrete physical space* be defined as space-for-a-particular-observer-instrument, and *concrete time* as time-for-a-particular-observer-instrument; these are the ordered manifold of distance and time intervals as measured by this observer-instrument. Any macroscopic thing in nature, however, can serve as an observer-instrument and so become the centre of perspective of the World-for-things. We are led to ask then, how the particular concrete space and time of one observer-instrument is related to the concrete space and time of another observer-instrument. Making the simplest assumption possible, Einstein found that the two spaces and times were related by a coordinate transformation belonging to the Poincaré (inhomogeneous Lorentz) group. In the later general theory of relativity, more complicated transformation groups were considered, but in all forms of the theory of relativity, the basic measure of distance is reduced to a time-like measurement on a standard clock.[5] The basic structure of space and time is expressed by the conservation of a scalar invariant (ds^2) under the continuous transformation group.

The Primacy of the Instrument

The new kind of mathematicised physics was not simply a rival to person-centered knowledge; for the person-centered world of observation is a condition for the existence of this knowledge. But beside the World-for-us, a new aspect of our World was discovered which was thing-centered; that is, centered on a part of nature itself. A measuring instrument in the new outlook played the part of an "observer-instrument" which "felt" and "observed" reality and "spoke" of its "experiences" to the scientist through the "language" of observable physical symbols; for the new science consciously took the point of view of an instrument immersed in nature. Col-

5. Cf., J. L. Synge, *Relativity: The General Theory* (Armsterdam: 1960) chap. III.

ours, sounds, etc., which constituted the World-for-us, took on a new symbolic character and became a "language" which "spoke" of the physical structure and interactions behind them. The scientist translated these into mathematical symbols, in which form they entered a mathematical theory which, as an intelligible whole, gave meaning to its terms.

How does an instrument—as a substitute observer immersed in nature itself—"speak" of its "experiences"? The measured property produces a macroscopic effect in the instrument; as, for example, a pointer reading on a scale, a "click" of a counter, or a track in a bubble chamber. This macroscopic effect is a material sign. A sign has a double reality: its *material reality* as a pointer, sound or bubble track, and an *intentional reality* proper to it as a sign, viz., as indicating something beyond its material reality; as, for example, a magnetic field intensity (the pointer), a charged particle (the "click"), or an elementary particle like the Omega-minus (the bubble track). Intentional reality is the mode of being of a sign as such. The act which confers intentional reality on a material sign, moreover, is not formally in the instrument, which is unconscious and devoid of the power of uttering significant signs, but formally in the mind of the scientist. The intentional reality of signs is the noematic correlate to an act of inquiring intelligence which is not content with what it sees, but looks for its explanation.

Instrumental data is more than a sign, it is a *symbol* of a physical property; for it *stands for* the property, as words in a language stand for the things they signify. Words in a human language, however, are conventional symbols; observable symbols in physics, such as the ones we have been considering, are *natural symbols*, since in a well-designed measuring process, they are uniquely determined by the interaction with the object. The "language" appropriate to an instrument when it "speaks" is a *"physicalistic language,"* in so far as it "translates" the hidden state of the object into a uniquely-determined sensible sign.[6]

The scientist, observing this sign, translates it in turn into a linguistic statement; as, for example, "This is the signature of an Omega-minus particle." This latter translation is not, however, a physicalistic one, for two reasons: (1) it describes more than the

6. Cf. K. Popper, *Conjectures and Refutations* (London: Routledge and Kegan Paul, 1963), pp. 296–297 for a discussion of the language of an instrument.

mere material reality of the data, i.e., it is *not* a mere *object language expression,* and 2) the data itself does not determine a unique linguistic statement. The statement "This is the signature of an Omega-minus particle" refers to a material object, viz., the signature, but only as a symbol of something which is not given as a material object, viz., the Omega-minus particle.[7] Moreover, the instrumental data does not determine *of itself* a unique linguistic statement; for it responds to the type of question formulated by the scientist. Of itself, the instrument is "dumb"; it waits to be questioned by the scientist, and the form of the question structures its response. For example, the data may evoke a mere description of its material reality (a bubble chamber track), or an explanation of its intentional reality (a signature of an Omega-minus particle), or an assertion or denial of a hypothesis (the Eightfold Way). The instrument then responds to the noetic intention of the scientist; it does not create it. It "speaks" only if "questioned"; and the structure of its response mirrors the structure of the "question asked" of it.

Many physicists, for want of a better theory of knowledge, adopt a form of psycho-physical parallelism which entails a unique physicalistic translation of physical events into linguistic statements.[8] It is clear from what has just been said that this is not a correct account of scientific knowledge or language.

Section II: The Language of Physics

Two Languages

The distinction and interrelation between observation and explanation in physics can also be explained from the point of view of language. A language is composed of a vocabulary, syntactic rules

7. The existence of the Omega-minus particle (mass, 1678; isotopic spin, 3/2; parity, plus) was predicted by a scheme of symmetry known as the "Eightfold Way." The prediction was made by Gell-Mann, Y Ne'eman, Salam and Ward. The discovery of the particle was announced by R. P. Schutt and his team at Brookhaven Laboratory in *Phys. Rev. Letters,* Feb. 24, xii (1964), 204–206.

8. For example, N. Bohr, J. von Neumann, E. Wigner, S. Watanabe and others. Cf., N. Bohr, *Atomic Theory and the Description of Nature* (Cambridge: 1961), pp. 24, 115–119.

governing the way words are connected into phrases and sentences, and a dictionary of correct usage governing the application of linguistic expressions (words, phrases, sentences, etc.) to their referents.

A language originates in a noetic viewpoint and is the instrumental expression of this viewpoint. The World described by a language is the noematic response to the originating noetic orientation. Now the noetic orientation of a physicist is twofold: 1) he works with, describes and observes pieces of apparatus and their behavior, in so far as these are parts of a *World-for-him,* and 2) he expresses what happens in nature from the perspective of an instrument, immersed in nature; the World to which he orients himself is one structured about things; it is a World-for-things.[9] Corresponding to this twofold noetic orientation of the physicist one expects to find, and one does in fact find, two languages: an *observation language* to express operational and observational concepts, and an *explanatory language* to express explanatory concepts.

It would seem to be at first sight plausible that the distinction should be caused by the existence of two kinds of referents for physical statements, viz., observable facts in their material reality and theoretical constructs (like electrons and Omega-minus particles) to explain these facts. One may be led uncritically to assume that the observable symbols are to be described by an observation language, while the theoretical constructs are to be described by a theoretical language. This indeed is commonly held by philosophers, as, e.g., by Wilfrid Sellars, Reichenbach, Carnap and others. We do not think it correct for the following reasons:

1) Even theoretical constructs, like electrons, are observable although only indirectly and by interpretation. This was implied in the content of Heisenberg's original insight that only *observables* should enter into the structure of physics. Observation language is not restricted, therefore, to the material sensible sign, but correct observation statements can be formulated even about theoretical entities. When a physicist observes, for example, a track in a bubble chamber, he does not merely study material marks, but he observes them as the signature of some particle, and his noetic intention is *through the observable signature to the particle.* Such noetic ac-

9. Or more correctly a *World-of-things-to-things-for-him,* as we pointed out in note I, p. 59 [of original work, Ed.]

tivity results in making particles and other theoretical constructs a part of the World-for-us; they take on the character of being given in perception—although of course not "given primordially in perception" [10]—but none the less rationally given by interpretation.

2) Moreover, the physical interaction between bodies given directly in perception can also be explained theoretically. One thinks of Eddington's two tables: one, the solidly perceived table; the other, a chaos of vibrating infinitesimal particles and constituted mostly of emptiness.[11] Precise analysis shows that no physical property or thing, in so far as they are a part of an explanatory physical context, is *per se* representable in sensibility. It is illegitimate to try to visualise the table as *explained,* for visualisation belongs to description and not to explanation. Whatever is visualised is a property-for-us; explanatory properties are not properties-for-us, but properties-for-things.[12] Hence, while the table has two sets of predicates each of which is correct, notice must be taken of the fact that they belong to different languages. It can be misleading to translate explanatory predicates into such typical terms as "chaos" and "emptiness." In our view both the accounts given above are correct; both tables are real, and they have one identical referent.

The difference between *observation language* and *explanatory language,* then, is not that they deal with different sets of referents, but that they consider the same set within different contexts. One considers them within the context of a World-for-us, while the other considers them within the context of a World-for-things. For this reason we prefer the pair of terms "observation language" and "explanatory language," for they do not suggest contrast and opposition to the same extent as the more usual alternative pair, viz., "observation language" and "theoretical language."

It may be objected that theoretical entities like elementary particles, etc., are ontologically on a different footing from objects like those given directly in perception, and that this distinction should be recognised linguistically by having one language for observed bodies and another for theoretical constructs like elementary parti-

---

10. Edmund Husserl, *Ideas* (London: Allen and Unwin, 1931), p. 52.
11. Sir Arthur Eddington, *The Nature of the Physical World* (Camb. Univ. Press, 1928), p. xiv.
12. Heelan, pp. 59, 61.

cles. This apparently is the opinion of W. Sellars, Carnap and many others. We answer that we too recognize that there is a difference in the ontological status of the two classes of objects, but in our view this difference does not constitute the real differences between the two languages. A language expresses a systematic whole, that is, a totality of referents related to one another in a particular way. "Observation language" and "explanatory language" represent, as we have said, different systematic totalities; but the classes of object to which they refer are both *real.* Their different ontological status results not from the supposition that one set is real while the other set is ideal (or purely mental), but from the fact that different kinds of criteria are used according to which their reality is known to us. In the case of bodies given in perception, it is the coherence of what is given with the rest of experience (not, of course, merely as a given coherence—for the dream-world too has its coherence— but as one whose coherence is critically and reasonably accepted). The criterion of a theoretical construct like an elementary particle, on the other hand, is more complex and involves 1) the ability to make virtually unconditioned judgements about the truth of a physical theory, and 2) the establishment of a connection between certain observable criteria and the reality of the physical entities which constitute the physical explanation. The same essential epistemological problem arises for all physical theories, whether one discusses the reality of mass, force and gravitation or the reality of mesons, neutrinos and Omega-particles. The only difference between the two sets of cases is that it is more difficult to satisfy the strategic criteria in the second set than in the first set.

Others, as we have said, have different views. We cite Wilfrid Sellars as an example.[13] He holds that the correspondence rules which form the "bridge" between the observation language and what he calls the theoretical language "appear in the material mode as statements to the effect that objects of the observational framework *do not really exist—there really are no such things.*" Such a view is consistent with the epistemological position that reality is constituted, or at least manifested, by the relation of exteriority be-

13. Wilfrid Sellars, "The Language of Theories" in *Current Issues in the Philosophy of Science,* ed. by G. Feigl and G. Maxwell (New York: Holt, Rinehart and Winston, 1961), p. 76. [This article is also reprinted in the present volume. Ed.]

tween a subject and a fully constituted phenomenal object. We are opposed to this view. For us, the criterion of a real thing is the *rationality* of the unconditioned affirmation of an object presented in sensation, whether directly or under its observable symbol, and understood whether in relation to ourselves (descriptively) or in relation to other things (explanatorily). This implies a meaning of the term 'reality' different from that of Wilfrid Sellars. It also implies a different criterion for the discernment of what is real.

Summary

This chapter puts into systematic form some of the clues to the nature of a physical theory, which have arisen during the course of the previous chapters. First of all, we treat of the two parts of a physical science: the formal mathematical theory whose function is to describe a World-for-things, and the experimental (operational and observational) part which makes this World-for-things, also a World-for-us. Secondly, we discuss the space and time of physics, distinguishing it from the space and time of perception. Thirdly, we describe the key concept of an observable symbol. Finally, we give an account of the two languages used by a physicist: the explanatory language and the observation language.

Bibliographical Essay

This bibliographical essay lists some selected books and periodical articles that either deal with the issues treated in the Introduction or supply helpful background material to the readings. Obvious reference works, such as encyclopedias and general histories of philosophy, are not listed. The listing follows the general order of topics treated in the Introduction. Whenever possible English translations and paperback versions are cited.

History of Science

For a factual background of science, two good, relatively brief surveys are: Stephen Mason, *A History of the Sciences*, rev. ed. (New York: Collier Bks., paperback, 1962) and A. Hall and M. Hall, *A Brief History of Science* (New York: Signet Bks., New American Library, paperback, 1964). The most comprehensive general survey is René Taton, ed., and A. Pomerans, trans., *History of Science*, 4 vols. (New York: Basic Books, 1963–1966). Each volume presents a collection of articles covering every aspect of scientific development with little or no discussion of philosophical issues. The accepted model of scientific growth through the gradual accretion of knowledge has been sharply criticized by Thomas Kuhn in *The Structure of Scientific Revolutions*, 2nd. ed. (Chicago: Univ. of Chicago Press, paperback, 1970). His own interpretation of conceptual revolutions in terms of scientific paradigms has in turn met severe criticism, e.g., Dudley Shapere, "The Structure of Scientific Revolutions," *Philosophical Review*, 73 (1964): 383–94. N. R. Hanson, *Patterns of Discovery* (Cambridge: Cambridge Univ. Press, 1958) counters the positivistic disregard of scientific discovery and conceptual innovation by showing how key scientific discoveries provided answers to pressing problems. Hanson's approach has been extended by Willard Humphreys, *Anomalies and Scientific Theories* (San Francisco: Freeman C., 1968) in showing how the dominant theories have specified the anomalies that require explanation.

Philosophical views condition scientific development in many ways. The influence of successively dominant views of the world as organism, as mechanism, and as a mystery unravelled by mathematics is treated by Stanley Jaki in the first three chapters of his *The Relevance of Physics* (Chicago: Univ. of Chicago Press, 1966). The most authoritative treatment of the mechanistic world view, a view that conditioned much of philosophy since Descartes, is E. J. Dijksterhuis, *The Mechanization of the World Picture*, trans. C. Dikshoorn (Oxford: Clarendon Press, 1961). C. F. von Weizsacker, *The Relevance of Science: Creation and Cosmogony* (New York: Harper & Row, 1964) is an illuminating, non-technical survey of cosmological theories and their influence on theology. The dialectical interaction between science and philosophy may often be seen more clearly through a detailed historical treatment of particular problems. Here the standard for competent, informed discussion has been set by Max Jammer in his three works: *Concepts of Force* (1962), *Concepts of Space* (1961), and *Concepts of Mass in Classical and Modern Physics* (1964)—all published in New York by Harper & Row. The growth of ideas concerning the nature of matter has been treated by A. G. van Melsen in *From Atomos to Atom* (Pittsburgh: Duquesne Univ. Press, 1953); and in S. Toulmin and J. Goodfield, *The Architecture of Matter* (New York: Harper & Row, paperback, 1966), theories of animate as well as inanimate matter are surveyed. The historical development of the concept of matter and its philosophical implications is the theme of a symposium presented in E. McMullin, ed., *The Concept of Matter* (Notre Dame, Ind.: Univ. of Notre Dame Press, 1963).

Particular Periods

One of the most scholarly and comprehensive accounts of ancient science through the Alexandrian period, George Sarton's *A History of Science*, 2 vols. (Cambridge: Harvard Univ. Press, 1952–1959) is somewhat marred by the author's rather intemperate attack on Plato and his general distrust of speculative philosophy. A good general survey is Marshall Clagett, *Greek Science in Antiquity* (London: Abelard-Schuman, 1957). The best general account of Aristotelian science is Friedrich Solmson, *Aristotle's System of the Physical World: A Comparison with His Predecessors* (Ithaca, N.Y.: Cornell Univ. Press, 1960). The criticisms that the neoplatonists of late antiquity directed against the Aristotelian idea of scientific explanation are summarized in S. Sambursky, *The Physical World of Late Antiquity* (London: Routledge & Kegan Paul, 1964).

The best general survey of medieval science is A. C. Crombie, *Medieval and Early Modern Science,* rev. ed., 2 vols. (Garden City, N.Y.: Doubleday, Anchor Books, paperback, 1959). A briefer book, James Weisheipl, O.P., *The Development of Physical Theory in the Middle Ages* (New York: Sheed & Ward, 1959) couples an interpretive survey with a defense of Aristotelian views on science. The historically significant transition from late medieval to early modern theories of motion is summarized in Marx Wartofsky, *Conceptual Foundations of Scientific Thought* (New York: Macmillan, 1968), Appendix A. His extensive bibliography, pp. 545–48, should be supplemented by W. Wallace, O.P., "The Enigma of Domingo de Soto: *Uniformiter difformis* and Falling Bodies in Late Medieval Physics," *Isis* 59 (1968), 384–401, which fills a gap on the immediate predecessors of Galileo.

Renaissance and early modern science has been extensively discussed and documented. A brief, lucid treatment of the changing physical ideas is I. B. Cohen, *The Birth of the New Physics* (Garden City, N.Y.: Doubleday, Anchor Books, paperback, 1960). A relatively non-technical, and occasionally misleading, general survey is Herbert Butterfield, *Origins of Modern Science, 1300–1800,* rev. ed. (New York: Macmillan, paperback, 1960). A more detailed treatment may be found in A. R. Hall, *The Scientific Revolution, 1500–1800* (Boston: Beacon Press, paperback, 1956). The changing conceptual background is extensively discussed in two related books: Marie Boas, *The Scientific Renaissance, 1450–1630* and A. R. Hall, *From Galileo to Newton,* which are volumes 2 and 3 of the series, "The Rise of Modern Science" (New York: W. Collins, 1962, 1963). E. A. Burtt's *The Metaphysical Foundations of Modern Physical Science* (Garden City, N.Y.: Doubleday, Anchor Books, paperback, 1954) has achieved the status of a modern classic in its discussion of the philosophical and theological influence of the scientific revolution. A different evaluation may be found in C. G. Gillispie, *The Edge of Objectivity: An Essay in the History of Scientific Ideas* (Princeton, N.J.: Princeton Univ. Press, 1960). Scientific developments induced changes in theories of scientific explanation, a topic treated in terms of representative thinkers in R. Blake, C. Ducasse, and E. Madden, *Theories of Scientific Method: The Renaissance through the Nineteenth Century* (Seattle: Univ. of Wash. Press, 1960) and in Joseph J. Kockelmans, ed., *Philosophy of Science: The Historical Background* (New York: Free Press, 1968).

Nineteenth-century science has not yet received the critical historical study it deserves. Apart from Taton, *History of Science,* vol. 3, one of the few general surveys is H. T. Pledge, *Science Since 1500* (New York: Harper & Row, paperback, 1959). Joseph Agassi, *Towards a Historiography of Science* (The Hague: Mouton, 1963), treats nineteenth-century physics, though it is not his central concern. Twentieth-century developments present a special problem because of the compartmentalization of the sciences and the technical nature of the new developments. We will simply list some works that view the different scientific revolutions from a philosophical perspective or that supply background material pertinent to the problem of realism. The earliest revolution was in mathematics. One of the simplest surveys of the new ideas is Stephen Barker, *Philosophy of Mathematics* (Englewood Cliffs, N.J.: Prentice-Hall, paperback, 1964). A clear exposition of the significance of logicism, formalism, and intuitionism is contained in S. Körner, *The Philosophy of Mathematics: An Introductory Survey* (London: Hutchinson Univ. Library, 1960). A good though often technical collection of representative readings is contained in P. Benacerraf and H. Putnam, *Philosophy of Mathematics: A Book of Readings* (Englewood Cliffs, N.J.: Prentice-Hall, 1964).

The special theory of relativity has occasioned extensive philosophical discussion. A very helpful reference work is *Special Relativity Theory: Selected Reprints,* available from the American Institute of Physics, 335 East 45th St., New York, N.Y., 10017. The opening article is a detailed, evaluative bibliography of books and articles on all aspects of the theory. A. Einstein's, *The Meaning of Relativity* (Princeton, N.J.: Princeton Univ. Press, 1955) is surprisingly easy to read in its discussion of the special theory. A. Einstein and L. Infeld, *The Evolution of Physics* (New York: Simon & Schuster, paperback, 1961) gives a non-technical account of the special theory's conceptual background. An excellent though somewhat advanced account of the wider significance of the theory is contained in G. J. Whitrow, *The Natural Philosophy of Time* (London: Thomas Nelson, 1961). The most ambitious philosophical reworking of this theory is A. Grünbaum, *Philosophical Problems of Space and Time* (New York: Knopf, 1963). His carefully reasoned arguments have served as the point of departure for discussions even by those who disagree with him. A sharp critique of his views may be found in H. Putnam, "An Examination of Grünbaum's Philosophy of Geometry," *Philosophy of Science: The Delaware Seminar*, ed. B. Baumrin, vol. 2 (New York: Interscience, 1963), pp. 205–55. A more balanced appraisal may be found in the panel discussion on Grünbaum's philosophy of science contained in *Philosophy of Science*, 36 (1969): 331–400, 429–32. Grünbaum's reply to this panel, contained in *Philosophy of Science*, 37 (1970): 469–588 is a rethinking of his earlier views. However, his writing style is probably too technical for anyone not already proficient in the field. An excellent general survey that covers both the physics and the philosophy of the space-time problem is Bas C. van Fraassen, *An Introduction to the Philosophy of Time and Space* (New York: Random House, 1970). See also E. MacKinnon, "Time and Contemporary Physics," *International Philosophical Quarterly*, 2 (1962): 428–57. There are various anthologies of articles on the problems of space and time: J. J. C. Smart, *Problems of Space and Time* (New York: Macmillan, paperback, 1964); J. T. Fraser, *The Voices of Time* (New York: Braziller, 1966); R. M. Gale, *The Philosophy of Time* (Garden City, N.Y.: Doubleday, Anchor Books, paperback, 1967). Each contains detailed guides to further reading. Because of its abstruseness, the general theory of relativity has not yet had the philosophical impact of the special theory. A clearly written, non-technical survey by one of the leading scientists in this area is Peter Bergmann, *The Riddle of Gravitation* (New York: Scribner, 1968). The best survey of cosmological theories resulting from the general theory is J. D. North, *The Measure of the Universe: A History of Modern Cosmology* (Oxford: Clarendon Press, 1965), a book that presupposes a good background in theoretical physics.

It was published too early to include such recent developments as quasars, pulsars, and cosmic black-body radiation.

Quantum theory is even more technical and more closely related to the problem of scientific realism. Two of the most accurate and readable non-technical surveys are Alexander Kompaneyets, *Basic Concepts in Quantum Mechanics,* trans. L. F. Landovitz (New York: Reinhold, paperback, 1966) and Victor Guillemin, *The Story of Quantum Mechanics* (New York: Scribner, 1968). Two works are especially recommended for the historical development of quantum ideas. The first, H. A. Boorse and L. Motz, *The World of the Atom,* 2 vols. (New York: Basic Books, 1966), is an extensive collection of original papers from the atomism of Lucretius to recent theories of fundamental particles. Each article is preceded by a brief, clarifying commentary. Chapters 66 and 71 are essentially philosophical discussions by Heisenberg and Bohr. The second, Max Jammer, *The Conceptual Development of Quantum Mechanics* (New York: McGraw-Hill, 1966), is a well written but highly technical account of the mixture of scientific and philosophical ideas that conditioned the development of quantum theory until 1930, when the orthodox framework was essentially settled. Bohr and Heisenberg, who created the 'Cophengan interpretation' of quantum theory, have explained its philosophical significance. Neils Bohr's views are developed in his *Atomic Theory and the Description of Nature* (Cambridge: Cambridge Univ. Press, 1934) and his "Discussions with Einstein on Epistemological Problems in Atomic Physics," *Albert Einstein, Philosopher-Scientist* (New York: Harper & Row, Torchbook, 1959), pp. 224–30. A clear summary of Bohr's views is presented by one of his associates, Aage Peterson, in "The Philosophy of Niels Bohr," *Bulletin of the Atomic Scientist,* 9 (Sept. 1963): 8–14. The best summary of Heisenberg's views is contained in his *Physics and Philosophy: The Revolution in Modern Science* (New York: Harper & Row, 1958). These views are discussed in detail in Patrick Heelan, S.J., *Quantum Mechanics and Objectivity: A Study of the Physical Philosophy of Werner Heisenberg* (The Hague: M. Nijhoff, 1965) and found wanting from the viewpoint of the critical Thomism discussed in my Introduction. The best contemporary survey of philosophical problems of quantum theory is contained in Robert G. Colodny, ed., *Paradigms and Paradoxes:*

The Philosophical Challenge of the Quantum Domain (Pittsburgh: Univ. of Pittsburgh Press, 1972).

A leader of the opposition to the Cophenhagan view, D. Bohm, has developed his own philosophy of nature in his *Causality and Chance in Modern Physics* (Princeton, N.J.: Van Nostrand, 1957). Even those who reject Bohm's interpretation of quantum mechanics in terms of hidden variables should find this a valuable study in the philosophy of nature. A good summary of the Cophenhagen view coupled with a philosophical criticism may be found in P. K. Feyerabend, "Problems of Microphysics," contained in *Frontiers of Science and Philosophy*, ed. R. Colodny (Pittsburgh: Univ. of Pittsburgh Press, 1962), pp. 189–283. A general survey presenting a somewhat Bergsonian and strongly anti-positivistic interpretation of contemporary physics is Milic Capek, *The Philosophical Impact of Contemporary Physics* (Princeton, N.J.: Van Nostrand, 1961). Recent interest in philosophical problems engendered by quantum mechanics has been concerned with the quantum theory of measurement. The best technical treatment of this problem is J. Jauch, E. Wigner, and M. Yanase, "Some Comments Concerning Measurement in Quantum Mechanics," *Il Nuovo Cimento*, 48 (1967): 144–51. A general survey may be found in R. Schlegel, *Completeness in Science* (New York: Appleton-Century-Crofts, 1967), chap. 10. Attempts to achieve an acceptable solution have induced extreme philosophical positions, e.g., psychophysical parallelism. See E. G. Wigner, *Symmetries and Reflections* (Bloomington: Indiana Univ. Press, 1967), especially chaps. 12–14.

The theory of evolution still occasions philosophical disputes that are related to the problem of scientific realism through the emergence-reductionism controversy. For a general background see: T. Dobzhansky, *Evolution, Genetics and Man* (New York: Wiley, 1957); Dobzhansky's non-technical summary of his views in "Scientific Explanation—Chance and Antichance in Organic Evolution" in *Philosophy of Science: The Delaware Seminar*, ed. B. Baumrin, vol. 1 (New York: Interscience, 1963), pp. 209–22; and G. G. Simpson, *This View of Life* (New York: Harcourt, Brace & World, 1964). Both Dobzhansky and Simpson reject attempts to subordinate evolutionary patterns of explanation to already established philosophical systems. An excellent, brief summary of contemporary scientific views is given in A. S. Romer, "Major Steps in Vertebrate Evolution," *Science*, 158 (Dec. 29, 1967): 1629–37. Recent developments in genetic theory are given a charming and lucid explanation in George and Muriel Beadle, *The Language of Life* (Garden City, N.Y.: Doubleday, 1966).

Much current research is concerned with the chemical evolution that preceded and led to biological evolution. For an excellent and relatively brief summary of these developments, see Cyril Ponnamperuma and Norman W. Gabel, "Current Status of Chemical Studies on the Origin of Life," *Space Life Sciences*, 1 (1968): 64–96. For more detailed treatments see Dean H. Kenyon and Gary Steinman, *Biochemical Predestination* (New York: McGraw-Hill, 1969) and Melvin Calvin, *Chemical Evolution* (New York: Oxford Univ. Press, 1969). Teilhard de Chardin, *The Phenomenon of Man*, trans. B. Wall (New York: Harper & Row, Torchbook, paperback, 1961) contains an original synthesis of evolutionary theory and Christian theology written in lyric prose. In spite of an army of followers, his position has won little critical support from scientists and philosophers. The outstanding objections are vehemently presented in P. B. Medewar's review in *Mind*, 60 (1961): 99–106. The opposite extreme, a mechanistic explanation of evolution, may be found in Dean Woolridge, *The Machinery of Life* (New York: McGraw-Hill, 1966). The underlying epistemological problem of what constitutes an explanatory rather than a descriptive account of evolution is treated in T. A. Goudge, *The Ascent of Life: A Philosophical Study of the Theory of Evolution* (London: Allen & Unwin, 1961).

Philosophy of Science

Introductory Works

Since positivistic interpretations have long dominated the philosophy of science, works in this spirit are considered first. Carl Hempel, *Philosophy of Natural Science* (Englewood Cliffs, N.J.: Prentice-Hall, 1966) is a lucid little summary concentrating on the deductive model of scientific explanation. P. Frank, *The Philosophy of Science* (Englewood Cliffs, N.J.: Prentice-Hall, 1956) is somewhat dated but closer to actual science than many more formal accounts. R. Carnap, *Philosophical Foundations of Physics* (New York: Basic

Books, 1966) stems from a seminar taught by Carnap and edited by M. Gardner. In many respects this work represents an unfortunate regression to Carnap's earlier views. J. Kemeny, *A Philosopher Looks at Science* (Princeton, N.J.: Van Nostrand, 1959) is a non-technical account of logical problems involved in scientific explanation.

V. E. Smith, *Science and Philosophy* (Milwaukee, Wisc.: Bruce, 1965) gives a brief survey of the development of the different sciences and outlines the views of almost every prominent philosopher of science before giving his own neo-Aristotelian account of science. Paul Durbin's *Philosophy of Science: An Introduction* (New York: McGraw-Hill, 1968) supplements a collection of articles and excerpts representing various schools of thought with a commentary generally reflecting a position of Thomistic realism. P. Caws, *The Philosophy of Science* (Princeton, N.J.: Van Nostrand, 1965) covers every aspect of the philosophy of science in a collection of short, pithy chapters and defends a non-scholastic version of immediate realism. S. Toulmin, *The Philosophy of Science* (New York: Harper & Row, 1953) stresses the role of models in scientific theories and gives a rather operationalist account of scientific explanation. A stimulating discussion of the debate concerning the role of models in scientific explanation may be found in M. B. Hesse, *Models and Analogies in Science* (New York: Sheed & Ward, 1963). A good general introductory text is J. J. C. Smart, *Between Science and Philosophy: An Introduction to the Philosophy of Science* (New York: Random House, 1968). Peter Medewar's small and easily readable *Introduction and Intuition in Scientific Thought* (Philadelphia: American Philosophical Society, 1969) is a reflective analysis of scientific method by an outstanding biologist. His basic concern is to explain the differences between what philosophers say scientific method should be and how the practicing scientist really functions.

Advanced Works

This section deals with general works on the philosophy of science, while the next section concentrates on more specialized studies. Three turn-of-the-century classics that have a continuing influence are: P. Duhem, *The Aim and Structure of Physical Theory*, trans.

P. Wiener (Princeton, N.J.: Princeton Univ. Press, 1954); H. Poincaré, *Science and Hypothesis* (1905; reprint ed., New York: Dover, 1952); E. Meyerson, *Identity and Reality*, trans. K. Lowenberg (orig. trans., 1930; reprint ed., New York: Dover, 1962). The most influential general textbook is probably E. Nagel, *The Structure of Science* (New York: Harcourt, Brace & World, 1961). This gives a detailed treatment of the major philosophical problems concerned with scientific explanation, and the natural and social sciences. Although a neopragmatic viewpoint is defended, differing opinions are treated in detail. M. Wartofsky, *Conceptual Foundations of Scientific Thought* (New York: Macmillan, 1968) is an extremely well informed, objective survey of current problems with all major opinions represented. His concluding bibliographical survey, pp. 489–548, is the best guide of its kind available. A. Pap, *An Introduction to the Philosophy of Science* (Glencoe, Ill.: Free Press, 1962) is a carefully reasoned account in the tradition of logical analysis that dismisses realism as a pseudoproblem. Two books reflecting realist interpretations of science by authors not affiliated with any school of philosophy are: L. O. Katsoff, *Physical Science and Physical Reality* (The Hague: M. Nijhoff, 1957) and D. Hawkins, *The Language of Nature* (New York: Doubleday, Anchor Books, paperback, 1967). Where Katsoff concentrates on the logical and linguistic problems of scientific explanation, Hawkins is more concerned with finding an order in nature, including human nature, and society. R. B. Braithwaite, *Scientific Explanation* (Cambridge: Cambridge Univ. Press, 1955) is a very formal account of the logic of scientific explanation in deductive, inductive, and statistical arguments. H. Margenau, *The Nature of Physical Reality* (New York: McGraw-Hill, 1950) is a neo-Kantian account stressing the constructural nature of theoretical knowledge. Margenau's mastery of theoretical physics, particularly quantum mechanics, greatly contributes to the value of this study. S. Körner, *Experience and Theory* (New York: Humanities Press, 1966) is also in the neo-Kantian tradition. A detailed criticism may be found in my "Epistemological Problems in the Philosophy of Science," *Review of Metaphysics*, 12 (1968): 113–137. Henry E. Kyburg, *Philosophy of Science: A Formal Approach* (New York: Macmillan, 1968) concentrates on the development of formal axiomatic systems and then seeks to reconstruct selected, and often trivial, portions of mathematics, physics, psychology, sociology, and biology as formal axiomatic systems. At the other extreme is another recent book, Peter Achinstein, *Concepts of Science: A Philosophical Analysis* (Baltimore: Johns Hopkins Press, 1968) Achinstein brings to bear on scientific language techniques of language analysis usually restricted to ordinary language. Thus, to explain how terms, theories, and models function in actual scientific usage rather than in logical reconstructions of scientific theories, he concentrates on such questions as, "When would we say that someone has a theory?"

Finally, there should be mentioned some works, too specialized or too difficult to serve as textbooks, by authors who have played a dominant role in the development of the philosophy of science. Bertrand Russell's views have undergone significant changes in the course of his long intellectual development. The most complete statement of his general views on the problems discussed here is contained in his *Human Knowledge: Its Scope and Limits* (New York: Simon & Schuster, paperback, 1962). Rudolf Carnap, long the leading logical positivist, summarizes his present position on almost every basic issue in his reply to his critics in P. A. Schilpp, ed., *The Philosophy of Rudolf Carnap* (La Salle, Ill.: Open Court, 1963), pp. 859–1053. See especially pp. 868–73 for his views on the problem of realism. Carl Hempel's *Aspects of Scientific Explanation* (Glencoe, Ill.: Free Press, paperback, 1970) contains a collection of his basic articles and concludes with a long account written with his usual clarity and objectivity, called "Aspects of Scientific Explanation," pp. 331–489, explaining his present views on the nature of scientific explanation and answering objections that Feyerabend, Scriven, and others have directed against his position. Hans Reichenbach, originally affiliated with the Vienna Circle, gradually became more of a physicalist who stressed the probabilistic aspects of knowledge. His views on knowledge and scientific explanation may be found in his *Experience and Prediction* (Chicago: Univ. of Chicago Press; New York: Phoenix, paperback, 1961). Although his explanation of quantum theory in terms of a three-valued logic had little influence, his views on space and time, contained in his *The Philosophy of Space and Time* (reprint ed., New York: Dover, 1958) and in his later, more difficult work, *The*

Direction of Time, edited posthumously by M. Reichenbach (Berkeley: Univ. of Cal. Press, 1956) have been influential.

The most influential opponent of logical positivism and a significant philosopher in his own right is Karl Popper. His basic views are explained in his *The Logic of Scientific Discovery* (New York: Wiley, Science Editions, paperback, 1961). His anti-inductivist account of scientific explanation is captured in the title of his collected essays, *Conjectures and Refutations* (London: Routledge & Kegan Paul, 1963). While never an anti-metaphysician, Popper has said surprisingly little about the problem of realism. This is undoubtedly due to his view of metaphysics as a study of general conceptual frameworks, a view clarified by one of his leading disciples, Joseph Agassi in "The Nature of Scientific Problems and Their Roots in Metaphysics" in M. Bunge, ed., *The Critical Approach to Science and Philosophy: In Honor of Karl B. Popper* (Glencoe, Ill.: Free Press, 1964), pp. 189–211. Mario Bunge's realistic interpretation of science may be found in his *Causality: The Place of the Causal Principle in Modern Science* (Cleveland: World Publishing Co., Meridian Bks., paperback 1963), while his epistemological and interpretative base is clarified in his *Metascientific Queries* (Springfield, Ill.: Charles C Thomas, 1959) and in *The Myth of Simplicity: Problems of Scientific Philosophy* (Englewood Cliffs, N.J.: Prentice-Hall, 1963). The interpretative basis for his views is developed in detail in his two-volume work, *Scientific Research* (New York: Springer-Verlag, 1967). This is undoubtedly the most detailed and sophisticated development of the type of immediate realism that was criticized and eventually rejected in my Introduction. Another work, concentrating more on the type of reasoning actually employed in scientific practice than on a logical reconstruction of scientific theories, is Paul Durbin, *Logic and Scientific Inquiry* (Milwaukee, Wisc.: Bruce, 1968).

Metaphysical Positions

Any grouping of philosophers into schools is more indicative than definitive. The classifications given here are simply intended to suggest a general orientation of contemporary philosophers with respect to the problem of realism.

Marxism

Until recently Marxist—or at least, Soviet—explanations of realism in general, and scientific realism in particular, tended to be predetermined by Lenin's strictures against empirio-criticism. For a summary of such positions see Gustav Wetter, *Dialectical Materialism: A Historical and Systematic Survey of Philosophy in the Soviet Union*, trans. P. Heath (London: Routledge & Kegan Paul, 1958) and Thomas Blakely, *Soviet Scholasticism* (Dordrecht, Holland: D. Reidl, 1961). *Soviet Studies in Philosophy*, vol. I, no. 2 (fall 1962) contains a series of articles on the philosophy of science by leading Russian philosophers. For a doctrine of realism and its relation to the problem of knowledge, see especially the articles by B. M. Kedrov, pp. 3–24, and by V. S. Tiokhtin, pp. 45–53. Newer trends may be found in P. V. Tavenec, ed., *Problems of the Logic of Scientific Knowledge*, trans. T. Blakely (Dordrecht, Holland: D. Reidl, 1970). While traditional Soviet realism is still maintained, the positions represented (except for the initial article by P. V. Tavenec and V. S. Svyrev) are generally more flexible and more open to Western influence than was possible in the past.

Thomism

Until recenty most Thomistic explanations of science presupposed immediate realism and used some version of the traditional doctrine of degrees of abstraction as a basis for a philosophical interpretation of scientific systems. The most original and influential work in this tradition is Jacques Maritain, *Distinguish to Unite or: The Degrees of Knowledge*, trans. G. Phelan (New York: Scribner, 1959). An excellent summary of his views may be found in Yves Simon's article "Maritain's Philosophy of the Sciences," *The Thomist*, 5 (1943): 85–102. Variant interpretations within this tradition of concept abstractionism may be found in Andrew van Melsen, *The Philosophy of Nature*, 2nd ed. (Pittsburgh: Dusquesne Univ. Press, 1954) and P. H. van Laer, *Philosophy of Science* (Pittsburgh: Dusquesne Univ. Press, 1956). For a book written in the scholastic tradition that does not rely on a doctrine of abstractionism, see Wolfgang Büchel, *Philosophische Probleme der Physik* (Freiberg, Germany: Herder, 1965), especially chap. 5. An informative summary of different approaches to the philosophy of science within the scholastic tradition may be found in William A. Wallace, O.P., "Towards a Definition of the Philosophy of Science," *Mélanges à la Mémoire de Charles De Koninck* (Quebec: Les Presses de l'Université Laval, 1968), pp. 465–85.

The tradition labeled 'transcendental Thomism' in the text stems from the pioneering efforts of Scheuer and especially Maréchal to redevelop Thomistic philosophy in a manner that is consonant with the critical requirements set by Kant. See Joseph Maréchal, S.J., *Le Point de départ de la Métaphysique*, vol. 5 of *Le Thomisme devant la Philosophie Critique*, 2nd ed. (Brussels: L'Edition Universelle, 1949); and Daniel Shine, S.J., *An Interior Metaphysics: The Philosophical Synthesis of Pierre Scheuer, S.J.* (Weston, Mass.: Weston College Press, 1966). European development of this approach is now available in English thanks to three translations published in New York by Herder & Herder in 1968: Karl Rahner, S.J., *Spirit in the World;* Emerich Coreth, S.J., *Metaphysics;* and Otto Muck, *The Transcendental Method.* A brief symposium on the validity of this movement may be found in *Continuum,* 6 (1968): 221–45. Bernard Lonergan, S.J., *Insight: A Study of Human Understanding* (New York: Philosophical Library, 1957), though in the same spirit, represents an independent, critical redevelopment of philosophy and a distinctive interpretation of the nature of scientific knowledge. For summaries and critical appraisals of Lonergan's thought see: *Spirit as Inquiry: Studies in Honor of Bernard Lonergan, S.J.,* ed. F. Crowe, S.J., (published both as vol. 2, no. 3 of *Continuum* and in book form by Herder & Herder in 1964); the series of articles by Edward MacKinnon, "Understanding According to Bernard J. F. Lonergan, S.J.," *The Thomist,* 28 (1964): 97–132, 338–72, 475–522; and David Tracy, *The Achievement of Bernard Lonergan* (New York: Herder & Herder, 1970). Although Tracy's account is more adulatory than critical it presents the most complete general survey of Lonergan's intellectual development. A detailed appraisal may be found in my review in *The Thomist,* 34 (1970): 654–63.

Analytic Pragmatism

Here, as in the Introduction, I use the label 'analytic pragmatism' as a blanket term covering recent philosophical developments that represent a fusion of elements drawn from the analytic, positivistic, and pragmatic traditions. Three significant works, roughly in order of increasing epistemological sophistication, are: J. J. C. Smart, *Philosophy and Scientific Realism* (London: Routledge & Kegan Paul, 1963), which defends a materialistic realism on the grounds that only physics and chemistry contain true scientific laws; W. V. O. Quine, *Word and Object* (Cambridge, Mass.: M.I.T. Press, 1960), which begins with ordinary language and the problem of meaning and argues toward a position in which the laws of logic play a definitive role in determining what should be accepted as real. An updated summary of his views may be found in his article "Ontological Relativity," *The Journal of Philosophy*, 65 (1968): 185–212. Finally, Wilfrid Sellars is developing a critical realism within the framework of a linguistic Kantianism. See his *Science, Perception and Reality* (New York: Humanities Press, 1963), especially chaps. 1 and 5; *Philosophical Perspectives* (Springfield, Ill.: Charles C Thomas, 1967); and *Science and Metaphysics: Variations on Kantian Themes* (New York: Humanities Press, 1968), which presents the epistemological basis for his doctrine of scientific realism; and "Scientific Realism or Irenic Instrumentalism" in R. Cohen and Marx Wartofsky, eds., *Boston Studies in the Philosophy of Science*, 2 (New York: Humanities Press, 1965), which explains the differences between his views and those of Nagel and Feyerabend and which also introduces the idea of the role of second order predicates adopted in my Introduction. Sellars, unfortunately, is notoriously difficult to read. It would be helpful to begin with a survey account of his views, such as, Richard Bernstein, "Sellars' Vision of Man-in-the-Universe," *Review of Metaphysics* 20 (1966): 115–43, 290–316 and my outline review of *Science and Metaphysics* in *The Philosophical Forum*, 1 (1969): 509–45.

For a general survey of logical analysis and its relation to ontic commitments, or to the type of realism implicit in the use of a logical framework, see Guido Küng, *Ontology and the Logistic*

Analysis of Language (Dordrecht, Holland: D. Reidl, 1967) and Gerard Radnitzky, *Contemporary Schools of Metascience* (Göteborg, Sweden: Akademiforlaget, 1968). Radnitzky supplements a highly critical appraisal of linguistic and logical analysis (vol. 1) with a highly laudatory appraisal of phenomenological analysis and hereneutics (vol. 2). Another author who attempts to relate logical analysis and phenomenology is Gustav Bergmann. For his unique fusion of logical analysis and platonic realism see his *Realism: A Critique of Brentano and Meinong* (Madison: Univ. of Wisc. Press, 1967), especially Book 1. The phenomenologists themselves have been more concerned with the grounding of science in transcendental subjectivity as constitutive than with the philosophy of science in the strict sense. In extending phenomenology as an interpretative base for science, most phenomenologists have been more concerned with psychology and the social sciences than with the natural sciences. However, the work that has been done has been collected by J. Kockelmans and T. Kisiel in their large anthology, *Phenomenology and the Natural Sciences* (Evanston: Northwestern Univ. Press, 1970). Here realism emerges as a problem especially in the discussions of Husserl (pp. 5–92) and Heidegger (pp. 147–204). In spite of their differences both agree that science seems to be objective because it does not radically reflect on its foundations. Philosophy is more radical and begins its critical work by suspending the natural viewpoint of science. In this perspective scientific realism is seen as unproblematic because unreflected. Only a reflective philosophy utilizing a phenomenological method can really raise the question of realism as a transcendental question. I hope to treat this problematic in more detail elsewhere.

Holistic Approaches

Here again we are using a classification that is more indicative than definitive. A. N. Whitehead's *Process and Reality* (New York: Harper & Row, paperback, 1960) is a major work that has influenced many metaphysicians, though it has had little influence in the philosophy of science. Three important contemporary works, each of which stresses the idea that in knowing, the grasp of a whole is prior to and more basic than an analysis into parts, are:

Michael Polanyi, *Personal Knowledge: Towards a Post Critical Philosophy* (Chicago: Univ. of Chicago Press, 1958); Errol E. Harris, *The Foundations of Metaphysics in Science* (New York: Humanities Press, 1965); and Susanne Langer, *Mind: An Essay on Human Feeling* (Baltimore: Johns Hopkins Press, 1967). While each emphasizes a different aspect of knowing (tacit knowledge for Polanyi, the grasp of an organized whole for Harris, and an intuitive grasp of 'acts' for Langer), each is led to oppose the ontological reductionism that is usually a feature of scientific realism.